HIGH TECH AMERICA

HIGH TECH AMERICA

The what, how, where, and why of the sunrise industries

Ann Markusen · Peter Hall
Amy Glasmeier

Boston
ALLEN & UNWIN
London   Sydney

© Ann Markusen, Peter Hall and Amy Glasmeier, 1986
This book is copyright under the Berne Convention
No reproduction without permission. All rights reserved.

**Allen & Unwin, Inc.,
8 Winchester Place, Winchester, Mass 01890, USA**

Allen & Unwin (Publishers) Ltd,
40 Museum Street, London WC1A 1LU, UK

Allen & Unwin (Publishers) Ltd,
Park Lane, Hemel Hempstead, Herts HP2 4TE, UK

Allen & Unwin (Australia) Ltd,
8 Napier Street, North Sydney, NSW 2060, Australia

First published in 1986

Library of Congress Cataloging-in-Publication Data

Markusen, Ann R.
　High tech America.
Bibliography: p.
Includes index.
1. High technology industries – United States.
I. Hall, Peter (Peter H.)　II. Glasmeier, Amy.
III. Title.
HC110.H53M373　1986　　338.4'767'0973　　86-17460
ISBN 0-04-338139-1 (alk. paper)

British Library Cataloguing in Publication Data

Markusen, Ann
　High tech America.
1. High technology industries – United States
I. Title　II. Hall, Peter, *1932-*
III. Glasmeier, Amy
338.4'76'0973　　HC110.H53
ISBN 0-04-338139-1

Austin Community College
Learning Resources Center

Set in 10 on 13 point Zapf Book Light by Getset (BTS) Ltd
and printed in Great Britain by Anchor Brendon Ltd, Tiptree, Essex.

Preface

AS WE SAY at the start of Chapter 1, there is a huge unslaked thirst for information about high technology industry. And this comes not merely from the academic community, which wants to understand where high tech jobs are being created, but even more so from decision makers who want to know how to attract high tech to their city or state. In this book, we have tried to serve the needs of both groups. Inevitably, we shall have failed: our academic peers will want more conceptual and technical detail, while local politicians and planners will want a good deal less. As far as possible, we have tried to keep our technical explanations short and understandable; and we have shifted the details into appendices. But we apologize in advance to those who may suffer mental dyspepsia on encountering words like entropy index, regression analysis or location quotient. Like most technicalities, they are hard to explain but they each contain a point which is both simple and important for what we want to say.

Readers should also bear in mind – indeed they will probably not need reminding – that the data on which we have based our study refer to the mid-1970s. We badly wanted to include a more up-to-date analysis; we even considered postponing publication until we could get the necessary data, which were just appearing as we sent the book to press. But any academic analysis of a contemporary problem suffers from the belated appearance of the data and from the time that then must be taken to process and analyze them. So we have left the book as it is.

There is a very important practical implication, stressed in Chapter 10: we cannot be sure that the practical policy implications still stand. Maybe defense is not so important a factor now (though, on the basis of what we know, we would doubt that); maybe basic university research counts more than, from our analysis, it did then. We cannot say – unless and until we come to repeat our work, which we still hope to do.

Acknowledgements

THE WORK REPORTED here was supported by grants to the Institute of Urban and Regional Development, University of California, Berkeley, by the National Science Foundation and the Office of Technology Assessment of the United States Congress. We wish to acknowledge their generous support. In addition, Northwestern University's Center for Urban Affairs and Policy Research supported a portion of the final writing up.

As Principal Investigators, Ann Markusen and Peter Hall designed the research agenda. Amy Glasmeier researched and wrote Chapter 2, on the definition of high technology industries, and Appendices 1, 2, and 3. Hall wrote Chapters 1, 8, and 10; Markusen wrote the rest; Hall oversaw preparation of the manuscript for publication. However, throughout the two years of our joint efforts, our ideas – about what high tech is, where it can be found, and why – mutually grew and changed. Many of the critical decisions, such as inclusion of industries, measurement of dispersion, and specification of the model, were made in concert. In addition to her participation in the continuing modification of the research design, Glasmeier also played a leading role in the construction and manipulation of the data base.

We want to acknowledge the assistance of a number of people, without whose aid the study could never have been completed: Carlos Davidson, Allen Sonafrank, and the staff of the Quantitative Anthropology Laboratory, University of California, Berkeley, for programming and analytical assistance; Henry Ruderman, for computer programming; Scott Campbell, for programming and help with the longitudinal analysis; Linda Wheaton and Vijaya Nagarajan, for data entry assistance; Aviva Lev-Ari, for statistical consultancy; Art Havenner, for consulting on the econometric work; Adrienne Morgan of the Department of Geography, University of California, Berkeley, who drew the maps; and the staff of the Institute of Urban and Regional Development at Berkeley – especially Elizabeth Prince and Maureen Jurkowski, who typed the final manuscript. At Northwestern University, special thanks to Margo Gordon, Director of the Center for Urban Affairs and Policy Research.

Finally, we owe a real debt to our copy editor at Allen & Unwin, Elizabeth O'Beirne-Ranelagh, for her meticulous attention to the manuscript.

Contents

Preface vii
Acknowledgements ix
List of tables xv
List of figures xvii

1 **What, how, where, and why** 1
   QUESTIONS FOR THEORY AND ANALYSIS

   Four basic questions 3
   Introducing theory 4
   Structure of the book: a summary 6

2 **What high tech means** 10
   THE PROBLEM OF DEFINITION

   Alternative high technology definitions 11
   Defining industry by occupation data 16
   Notes 22

3 **How high tech grows** 24
   GAINERS AND LOSERS IN PLANTS AND JOBS

   Recent high tech job and plant growth 25
   Note 39

4 **Why high tech grows** 40
   THE PRODUCT–PROFIT CYCLE

   Stages in high tech evolution 40
   Military-related sectors: a special case 45
   Grouping high tech sectors by stage 47
   Notes 57

5 **How high tech organizes** 59
   COMPETITION, MONOPOLY, AND THE PROFIT CYCLE

   Profit cycle stages and market power 66

	Market power in contemporary fast-growing industries	68
	Market power in military-related sectors	68
	Summary	70

6 How high tech clusters — 71
GEOGRAPHICAL CONCENTRATION AND DISPERSAL

	Some theory: forces for agglomeration and dispersal	72
	The locational data base	79
	Geographical concentration and dispersion	80
	The geographical redistribution of high tech industry	88
	Some unresolved questions	93
	Summary	94
	Notes	96

7 Where high tech locates — 97
REGIONAL AND URBAN DISTRIBUTION PATTERNS

	Regional high tech agglomerations: the core high tech states	98
	The differing high tech base of the core states	103
	Agglomeration of high tech in regional centers	105
	The regional dispersion of high tech industries within major agglomerations	110
	Metropolitan high tech agglomerations	113
	Summary	129
	Notes	131

8 Why high tech locates — 132
TOWARD A THEORY OF LOCATION

	Traditional location theory	133
	The importance of innovation	136
	The geography of research and development	139
	Some conceptual issues	142
	Note	143

9 Where and why high tech locates — 144
APPLYING THEORY TO DATA

	A model of high tech job and plant location	145
	Model formulation	150
	Testing the model for high tech location	154
	Testing the model for high tech change	158
	Disaggregating the model	160
	Summary	167
	Notes	168

10 What policy could do   170
IMPLICATIONS FOR NATIONAL, STATE AND LOCAL POLICY MAKERS

Some policy-relevant findings	170
Implications for policy	176
Notes	181

APPENDIX 1
Sources of occupational data: Census versus OES   182

Identifying high technology industries by occupational mix	183
Notes	185

APPENDIX 2
The Census of Manufactures plant location data base   186

The data base and previous users	186
Estimating the means for 'high technology' industries	188
Test of mean estimates	190
Cautions on the interpretation of employment data	190
Notes	191

APPENDIX 3
The independent variables   193

Airport access	193
Black population 1970	193
Business services	194
Climate index	194
Defense spending per capita	194
Educational options	195
Fortune 500 headquarters	194
Freeway density	194
Housing price	195
Unemployment rate 1977	195
Unionization rate	195
Wage rate	195
University R&D	195

APPENDIX 4
Specifying the model   196
Notes	200

APPENDIX 5

Disaggregating the regression analysis by SMSA size classes — 201

Large metropolitan areas	201
Medium-sized metropolitan areas	203
Smaller metropolitan areas	204
Notes	211

References — 212

Index — 219

List of tables

2.1	A product-sophistication definition of high technology: Massachusetts	12
2.2	Growth rate definition of high tech: USA	13
2.3	A combination definition of high technology: California	14
2.4	An R&D-based definition of high tech: USA	15
2.5	The 29 high tech sectors and their occupational mix	18
2.6	The 100 high tech industries	20
3.1	Jobs and plants in high tech industries, 1972–81	25
3.2	Top 15 industries ranked by employment level and plant incidence, 1981	30
3.3	Top 30 industries ranked by absolute job gains, 1972–81	31
3.4	Job growth rates for high tech industries, 1972–81	33
3.5	Industries ranked by percentage job gains, 1972–7 and 1977–81	38
4.1	High tech sectors classified by profit cycle location, 1947–81	50
5.1	Concentration ratios for high tech industries, 1977	60
5.2	Highly oligopolized high tech industries	64
5.3	Moderately oligopolized high tech industries	65
5.4	Competitive high tech industries	66
5.5	Concentration ratios in defense-related industries	69
6.1	Plant sites of San Jose-based firms, by phase of production, 1980	78
6.2	Highly dispersed high tech industries	80
6.3	Highly concentrated high tech industries	81
6.4	Entropy indices: a measure of spatial dispersion of high technology industries, 1972 and 1977	82
6.5	County and city locations entropy index and coefficient of redistribution, 1977	90
6.6	Ranking redistributive high tech sectors	93
7.1	Leading high tech states, 1977	99
7.2	High tech state agglomerations, 1977	101
7.3	Core state high tech agglomeration, 1972–7	106
7.4	Plant and job change by state and region, 1972–7	107

7.5	High tech state agglomerations: plant and employment change, 1972-7	110
7.6	Top ranked metropolitan areas, job and plant levels and change, 1972–7	114
9.1	Factors associated with metropolitan high tech job and plant location	155
9.2	Location factors associated with high tech employment, 1977, by profit cycle group	164
A2.1	Data sources used to calculate missing data in employment	190
A4.1	The 13 exogenous variables: correlation matrix	198
A5.1	SMSAs arrayed by size, 1977	205

List of figures

4.1	Employment change, innovating industries, 1947–81	49
4.2	Employment change, market penetrating industries, 1947–81	52
4.3	Employment change, volatile industries, 1947–81	53
4.4	Employment change, market saturated mature industries, 1947–81	55
4.5	Employment change, rationalizing (declining) industries, 1947–81	56
7.1	High tech employment: location quotients, 1977	100
7.2	The five core areas and their fringes	102
7.3	Greater Los Angeles	118
7.4	Chesapeake to Hudson	120
7.5	Greater San Francisco Bay	121
7.6	Old New England	123
7.7	Lower Lake Michigan	124
7.8	Florida	126
7.9	The Colorado front range	127
7.10	The Salt Lake corridor	129

1

What, how, where, and why

Questions for theory and analysis

IN THE MID-1980s, high technology has become the economic Holy Grail. In the countries of the advanced industrial world, it is hailed as the antidote to the decline of the smokestack industries of the 19th and early 20th centuries. In the newly industrializing countries of Latin America and East Asia, it is seen as the key to rapid economic catch-up. National political leaders commit their countries to it. Local elected officials seek to attract it. The new centers of high tech industry – California's Silicon Valley, Massachusetts' Highway 128, England's M4 Corridor, Scotland's Silicon Glen – exert an almost hypnotic hold over media people, and over their readers and viewers.

Nowhere is this more clearly evident than in the United States. In March 1985, the *New York Times* published a special Sunday supplement on the theme *High technology employment outlook*; the opening feature was entitled *Every region gets into the technology act*. Earlier, when we published an obscure working paper in the course of our present research, the *Wall Street Journal* picked up the fact; the resulting deluge of inquiries jammed our telephones for days. People are very thirsty for information about high tech: about the kinds of jobs it creates, about how many of them there are, and above all – for local decision makers – about where these jobs are, and why.

This thirst for information has gone largely unslaked. The press reports consist largely of anecdotal snippets. From academia has come a stream of contributions on different aspects of the high tech phenomenon, but they are all partial, dealing with single industries, single areas or single aspects. The California Commission on Industrial Innovation has commissioned and published a useful series of studies on the present state and future prospects of high tech industries in that state – computer software, photovoltaics, robotics, biotechnology, and others – which set their analyses within a broader nationwide framework (for summaries, see Hall & Markusen 1985). There are studies of the genesis and growth of Silicon Valley, which offer deep insights into the processes that generate high tech industries at critical moments in history (Saxenian 1981, 1985, Rogers & Larsen 1984). And there is a whole series of contributions from Edward Malecki, which have given a well-rounded picture of the geography of research and development in the United States (Malecki 1980a, 1980b, 1980c, 1981a, 1981b).

These studies give us tantalizing partial glimpses of the high tech landscape. We know, for instance, that high tech industries demonstrate astonishing rates of early growth – as for instance computer software (Hall *et al.* 1985). We understand that in the genesis of Silicon Valley the critical role was played by the juxtaposition of an institution (Stanford University), an entrepreneurial individual (Professor Frederick Terman), and historically determined demand (from the Pentagon, in the Cold War–Korean War period; Saxenian 1985). We appreciate that there is a critical difference between industrial research and development (R&D), much of it concentrated in headquarters cities of major corporations within the older manufacturing belt, and federal R&D, much of it based in newer Sunbelt centers and most of it defense-based (Malecki 1981a).

But the overall picture is lacking. There has been no systematic account of what high tech is, what kinds of industry make it up and how they are organized, where they are located and why, and how these patterns are related to the current discussion of the new international division of labor. It was this dearth of basic information – and thus of any satisfactory explanation of the high tech phenomenon – that led us in autumn, 1982 to embark on the study which we report here.

Four basic questions

In this study, we are concerned with four basic questions. In order of their appearance:

What? What exactly do we mean by the elusive term, 'high tech'? What criteria have others used to define an industry as a high technology industry? What kinds of problems do these criteria present, and can we find a better criterion? What is the resulting list of high tech industries?

How? How are high tech industries, thus defined, performing? Are they (popular stereotype) growing rapidly, adding plenty of new jobs? Or are there any high tech laggards? How is high tech industry organized? Is it characterized by a multitude of small plants, or is it becoming dominated by a few giant units? How is it distributed over the face of the United States? Is it, as many supposed, concentrated in a few favored locations: in Silicon Valley and its competitors? Or does it diffuse fairly widely from state to state and from city to city?

Where? Following that question up and concentrating now on the actual locations: where exactly does high tech industry locate? And – perhaps a question with a different answer – where is it growing? Where are the places with the most new factories, and – again, possibly a question with different answer – the places with the most new jobs?

Why? This question parallels and underlies all the others, running from chapter to chapter throughout the book. For we are concerned not to produce a mere almanac of high tech facts, but also, more importantly, to try to explain these facts. We ask: why should we distinguish some industries as high tech, and others as not? Why should high tech industries grow faster than the average, yet some grow faster than others? Why should we find some highly concentrated in a few firms and plants, others widely distributed among many competing companies? Why are some spatially concentrated in just a few locations, while others are widely distributed across the country? Why are high tech industries located where they are, and why are they growing faster in some places than in others?

Introducing theory

Because we are centrally concerned with these *why?* questions, we need to introduce some theory. This book relies very heavily on two important theoretical insights, derived from the literature in industrial economics and economic geography. We summarize them briefly now; later, in the appropriate places, we shall need to treat them in more detail.

The first is called *product cycle–profit cycle theory*. First developed by the economist Raymond Vernon (Vernon 1966), it depends on earlier theoretical insights developed especially by Joseph Schumpeter (1911, 1939). It suggests that products, and the industries that produce them, go through a cycle from youth to old age. All products, and all industries, have been at some time new, arising from technological innovation; thus, it can be argued, every industry was once high tech. At this early stage, the industry is typically characterized by a host of new small firms; they tend to cluster together in one or two areas, in order to enjoy the economies of agglomeration.

Later, as the industry matures, many of these early entrants drop out of the competitive race; more and more, the industry is dominated by a few successful firms, rapidly increasing in size. At this stage, too, *product* innovation – the development of a new product – becomes less important than *process* innovation – the constant search for more efficient production methods, especially by substituting capital for labor. Thus the industry, which at first generated plenty of new jobs in line with increasing output, enters a phase of jobless growth; and eventually, as market saturation is reached, actual employment decline. In this search for lower-cost production, the big monopolistic company will tend to desert the original center of production for other locations – in other regions, and even in other countries – where costs, especially of labor, are lower. Thus, paradoxically, as the industry becomes *organizationally* more concentrated, it becomes *geographically* less so.

The product cycle is also a *profit cycle* (Markusen 1985a). Early entrants to the industry may enjoy windfall profits from the rapid market growth of the product, but fierce competition, including many bankruptcies, sharply reduces the average rate of profit. Later, in the stage of maturity, the semi-monopolistic big corporation will achieve above-normal profits, as a result of eliminating competition and pursuing product differentiation. Finally, in old age the rate of profit will drop as the product reaches a stage of market saturation. Now, faced with a contracting market – and often with competition from newly industrializing countries which are able to imitate the technology, as well as from substitute products – companies are forced to pursue a policy of rationalization, with closures, layoffs and overseas relocation (Massey & Meegan 1982, Bluestone & Harrison 1982).

In the first half of this book (Chs. 3–6), we use product–profit cycle theory as a basic explanatory device. Through it, we seek to show why high tech plants grow faster than average (though some grow faster than others); why in some high tech industries output is spread across many firms while in others it is concentrated in a few; and, further, why some are concentrated in a few areas while others are much more widely diffused. In other words, in this half of the book we view today's high tech industries as arrayed on a continuum of evolution – albeit, by definition, clustered toward the leading edge of that continuum.

In the second half of the book (Chs. 7–9), we make use of a different though related set of theoretical insights, derived from *location theory*. Traditional industrial location theory, as developed by such pioneers as Alfred Weber (Weber 1929) and Edgar Hoover (Hoover 1948), relied heavily on the pulls of factors of production such as materials, markets and labor supplies, reinforced by economies of agglomeration. But, for today's high tech industries, only labor and agglomeration among these factors seem to be very important. High tech, we hypothesize, will be drawn to areas that are attractive to scarce, highly skilled professional and managerial talent – or, alternatively, to areas that offer a pool of weakly organized, poorly paid assembly workers. It will also cluster in areas that offer a well-developed industrial infrastructure of transportation facilities and business services.

But we need to go beyond traditional location theory, and to draw on insights from the product–profit cycle concept. If high tech industries are inherently innovative, then, we might surmise, they should flourish in areas that have the right business climate. If, further, their innovative character derives from fundamental scientific research and its development, then they should develop in places with a strong research tradition. Building partly on elements from traditional theory, and partly on notions derived from a cyclical theory of innovation, we build an eclectic explanation of high tech location, in which a number of different factors are embedded in a model that can be empirically tested. Finally, we try to combine the two theoretical models, so as to explain why different location factors come into prominence at different stages of an industry's evolution – one set dominating in youth, another in old age.

Structure of the book
a summary

The structure of the rest of this book follows directly from the four questions that are asked, and from the resulting interrelationship of theory building and data analysis.

Chapter 2 asks the basic *what?* question: what is high tech? It looks at alternative ways of defining a high tech industry. Rejecting a number of these, it argues that the best definition is one based on the human capital embodied in an industry, in the form of a high concentration of skilled scientists, engineers and technicians. On this basis, it emerges with a list of 29 industrial sectors based on the three-digit Standard Industrial Classification (SIC) of the United States Census of Manufactures, which disaggregate into 100 four-digit industries. These form the subject-matter of the analyses in the remainder of the book.

Chapter 3 asks the basic *how?* questions about high tech industrial performance. Using Census of Manufactures and other data, it shows that indeed high technology industries did provide over 1 million new jobs in the decade 1972–81: nearly nine in ten of all new manufacturing jobs in this period. It also shows that much of this growth came from the industries that everyone would conventionally expect: computing, radio and television communications, electronic components, semiconductors. But there is a less expected finding: no fewer than 33 of the 100 high tech industries actually lost jobs, and 19 others grew at a lower rate than the average for manufacturing generally.

Chapter 4 turns to the related *why?* question. It seeks to explain these variations in terms of the different stages that industries had reached in the product–profit cycle. It finds that the explanation generally holds good, but that some industries – especially those that are highly defense-dependent – constitute a special case.

Chapter 5 extends the list of *how?* questions. It asks how high tech industries are organized, and, in particular, whether production comes from many firms or from a few. The answer varies greatly from industry to industry. Some are highly oligopolistic, some are highly competitive. Again, this chapter interprets these results in terms of product–profit cycle theory, with mixed results: although for many industries it appears true that market power comes with maturity, there are several exceptions to the rule.

Chapter 6 continues to explore these *how?* questions with the aid of product–profit cycle theory. Turning to location, it asks: how concentrated, or how dispersed, are the high tech industries? It finds much variation, related not to the size of the industry, but rather to the type of product, the nature of the major consumer, and the dependency on natural resources for inputs; the highly innovative, fast-growing industries, which most people would regard as the archetypal high tech group, are moderately dispersed. But one finding stands out: there is a general tendency, with only few exceptions, for high tech industries to disperse over time. This might seem to be in line with what product–profit cycle theory would suggest; but testing shows that it is supported only in part.

Chapter 7 marks a major break in the book, which now turns specifically to the *where?* question. This, the first of a group of three chapters, concentrates on empirical analysis. It displays data, partly at the level of states and partly at the level of Standard Metropolitan Statistical Areas (SMSAs), which show where high tech industry is located and where it is growing. At the state level, it reveals a striking pattern dominated by five major high tech cores and five smaller cores, surrounded by penumbrae which are influenced, in greater or lesser measure, by proximity to the cores. This analysis disproves the conventional notion that high tech is concentrated in the Sunbelt, for two of the five major cores are found in the so-called Frostbelt; but the stereotype is true in one respect, that all the cores except one (Chicago) are located outside the Midwest industrial heartland that runs from Buffalo to St. Louis and Milwaukee.

Particularly notable, in this analysis, is the fact that the core high tech regions seem to be deconcentrating outward, into their surrounding fringe states, to some degree – though not all to the same degree. This pattern of dispersion also emerges at the finer-grained SMSA level of analysis. It can be seen that a number of the cores consist of complexes of adjacent SMSAs, in which the general tendency is for outward deconcentration from the large, central SMSA to fringe metropolitan areas; a pattern traceable for the Los Angeles, San Francisco, Middle Atlantic and Florida groups of SMSAs. At the same time, some of the most dramatic high tech job gains prove to have been made by small, remote metropolitan areas in the interior of the country, far from major urban agglomerations.

Chapter 8, balancing the empirical emphasis of Chapter 7, now concentrates entirely on the theory that throws light on the *why?* question. It looks at traditional Weberian location theory, and finds only some elements of it – particularly labor supply and agglomeration economies – to be particularly relevant to an analysis of high tech. It then turns to insights derived from cyclical theories of innovation and change, and concludes that some elements – business climate and scientific research – appear especially important for an understanding of high tech location. These elements form principal building-blocks for the construction of hypotheses about high tech location, which can be tested empirically.

Chapter 9 is concerned with these empirical tests of the *where?* theories. Thirteen possible explanatory factors, derived from the theoretical discussion in Chapter 8, are tested in a multiple regression model, using SMSA-level data for high tech location and locational change for the years 1972 and 1977. The results are statistically satisfactory in explaining location patterns, less so in explaining the changes over time. Perhaps the most interesting result of all is that some of the factors traditionally thought most important in attracting high tech industries, such as federally sponsored scientific research in universities, are in fact relatively insignificant; overall, the four most consistently significant factors are defense spending, range of educational options, presence of business services, and absence of black population. But the chapter ends with an important warning: these relationships are merely statistical in character, and they do not demonstrate that a causal relationship also exists.

Chapter 10 finally returns to the *what?* question. It discusses the findings of the study from the point of view of public policy, asking what policy makers – at national, state and local levels – could or should do to influence the patterns of future high tech growth and change. It extracts the most important policy-relevant findings from the study. High tech does create jobs; more than 1 million in a decade, in a whole range of industries. It is on the move; dispersing outward from a few high tech core regions to their peripheries, and within these regions from big-city metropolitan areas to smaller suburban-type ones. It has by no means deserted the older industrial regions and cities. But it benefits many small places remote from the major cities, which score some of the most spectacular high tech growth rates. Its location patterns are explicable – and therefore susceptible to policy; high tech industry is found to be highly sensitive to amenity factors, to accessibility advantages, and to agglomeration economies.

The chapter concludes with some very tentative policy recommendations. (It needs to be emphasized that they stem from the experience of the period 1972–7.) The first group is directed at local policy makers. One is that almost any place can join in the competition to attract high tech; though some are better-placed and others worse-placed, and though it takes time to build the base, nowhere is so hopeless it is out of the race. A second is that, clearly, defense spending is now a key variable and likely to remain so in the immediate future. A third concerns the mysterious missing variable of the study: fundamental research in universities and major research centers, which was expected to prove significant but failed to do so. Building high tech on a major research university, it seems, is a myth without foundation – at least on the basis of experience down to the mid-1970s. A fourth conclusion offers a word of caution. The future has seldom been like the recent past, and is unlikely to be so now. The factors that affect high tech growth in the next decades may be very different from those that dominated in the past. But the result is not likely to be completely different, for there is now so much inertia in the high tech map. It may therefore be a good policy for each local area to look to its particular variety of high tech strength, and seek to develop that in order to meet new competitive challenges.

A second group of recommendations is for national policymakers. Since political ideology must enter here, we offer Republican and Democrat scenarios of high-tech policy – and, since Democrat thinking proves to be divided, we suggest two alternative paths.

2
What high tech means

The problem of definition

BEFORE WE EVEN start to ask any other questions about high tech, we need to ask what high tech is. This might seem a simple, even simplistic, question. But it is not, because high tech is a portmanteau word that means many different things to different people. For state and local economic development planners, it means emerging growth industries that may provide the solution to high unemployment. For industry, it means new products and new, often labor-saving, production processes. In political circles, it means the promise of rejuvenation of America's competitive "edge." In academia and think tanks, it refers to more esoteric forms of research and development. Thus 'high tech' is a term with many meanings, because different people expect it to perform so many different roles: to provide jobs, to improve the productivity of the nation's industry, to protect and maintain America's economic competitiveness, and to produce new socially useful products and mechanisms which will raise our standard of living.

Yet in the midst of all these multiple meanings, the strange fact is that we lack a standard definition of what high tech is (Riche *et al.* 1983). In this chapter we critically review attempts at definition, and then present our own working definition. After reviewing the alternatives, we choose to specify high technology industries on the basis of the degree of sophistication and competence embodied in technical occupations – that is, on the human capital component of the labor process.[1]

Because our research focuses on the economic development potential of high technology industries, our definition does not automatically include those industries which consume high tech products and processes. For example, the adoption of robotics in the auto industry will not result in the reclassification of that industry as high tech. This failure to include all sectors affected by new technologies is regrettable, in that much of the net job *loss* associated with high tech innovation is thus not captured. If anything, our subsequent results are biased toward an optimistic assessment of high tech job potential and probably understate the profound geographical shifts that have occurred. The uses of new technologies have strongly shaped interregional development patterns, but documentation of this 'derived development effect' was beyond the bounds of our research design.

Alternative high technology definitions

The literature offers three frequently cited measures of high technology industries: (1) the perceived degree of technical sophistication of the product produced by an industry; (2) the rate of growth in employment within the sector; and (3) research and development expenditures as a percentage of sales. All of these measures lack operational precision.

Product sophistication

The Massachusetts Division of Employment Security (MDES) developed this measure on the basis of industries classified in the Standard Industrial Classification (SIC) Manual as cited in Vinson and Harrington 1979. Essentially, MDES staff analyzed the SIC manual and industries were selected on the basis of perceived product sophistication. Their list of 20 industries, shown in Table 2.1, includes several non-manufacturing sectors, which is a virtue, but does not offer a working criterion for sorting high tech industries from 'low tech.'

Growth in employment

A second and more widely used notion of high tech industries is based on industry employment growth rates. As seen in Table 2.2, according to the

Table 2.1 A product-sophistication definition of high technology: Massachusetts.

SIC	Industry	Employment 1978
281	industrial inorganic chemicals	1,272
282	plastics & synthetic resins	5,582
283	drugs	2,463
351	engines & turbines	5,817
357	office computing machines	32,430
361	electrical transmission equipment	11,891
362	electrical industrial apparatus	3,112
366	communication equipment	27,609
367	electronic components & assembly	40,555
372	aircraft & parts	9,229
376	space vehicles & guided missiles	12,438
381	engineering, laboratory instruments & scientific instruments	3,308
382	measuring & controlling instruments	19,325
383	optical instruments & lenses	6,435
385	photographic equipment	17,866
737	computer programming services	10,259
7391, 7397	commercial R&D laboratories	8,677
7392	business management & consulting services	9,003
891	engineering & architectural services	24,688
892	non-profit educational, scientific, & research organizations	7,723
Total high tech		259,682
Total Massachusetts private employment		2,132,695

Source: Vinson & Harrington (1979), p. 12.

National Science Foundation (NSF), between 1965 and 1977 employment in petroleum, chemicals, electrical equipment, and machinery increased on average by 18.4%, while employment in more traditional industries – textiles, food and kindred products, and stone, clay and glass – grew by only 2.4% over the same period. One simple way of choosing sectors is to define all those with an excess of job growth above the manufacturing average as "high tech," in this case all those sectors down as far as printing and publishing (Technical Marketing Associates 1979).

The problems with this approach are several. Sectors not apparently producing a high tech product, such as furniture and fixtures or rubber and plastics, are included. Indeed, the success of a sector in creating jobs may be attributable to many other factors besides technological sophistication. Employment growth as a measure of high tech would also exclude industries that are on the cutting edge of research, such as

What high tech means

Table 2.2 Growth rate definition of high tech: USA.

SIC	Industry	Percent total employment change (1965–77)
30	rubber & plastic products	43.4
38	scientific instruments	35.5
35	machinery	26.0
25	furniture & fixtures	18.4
36	electrical equipment	16.6
28	chemicals	16.5
29	petroleum refining	14.5
34	fabricated metals	14.4
27	printing & publishing	13.3
26	paper & products	9.4
22	textiles	6.1
24	lumber & wood products	5.8
32	stone, clay & glass	3.8
37	transportation equipment	3.2
39	miscellaneous manufacturing	−0.2
20	food & kindred products	−2.1
23	apparel	−4.8
33	primary metals	−7.4
31	leather products	−25.1
	Two-digit industry average	10.8

Source: Technical Marketing Associates (1979).

biotechnology, which have yet to produce widely commercialized products and attendant employment growth. Depending on the time span, many defense-related industries would also be eliminated because of the sporadic nature of that industry's growth, while on other measures, such as the degree of product sophistication and R&D, aerospace and defense industries would score very high indeed. Finally, this criterion would produce different sets over time; for instance, a fast-growing sector of the 1960s would drop out as employment growth levelled out in the subsequent decades.

Fundamentally, this approach contains a serious conceptual error. It is based on the tautological notion that since high tech industries are anticipated to be rapid net job generators, they can be *defined* by this performance. But to do so is to fail to acknowledge other sources of job growth, such as relative price changes, resource exhaustion, import substitution, or a successful marketing effort. It would also deny the possibility that high tech sectors might *not* be high growth performers. In other words, evidence on job creation could not then be used as a test of high tech economic development payoffs.

Table 2.3 A combination definition of high technology: California.

Industry	1980 employment	1990 employment	Annual growth rate (%)
biotechnology	2,000	9,000	18
photovoltaics	1,000	4,000–10,000	15–26
robotics/computer aided manufacturing	1,000	5,000–10,000	18–26
computer software and data processing services	43,300	128,000	11.5
computers and peripherals	97,100	173,000	6.0
electronics components	123,100	218,384	5.9
aircraft and space	207,500	217,600	0.5
instruments	98,600	132,500	3.0
communication equipment	130,100	166,600	2.5
Total high technology	703,700	1,065,080	4.3
Total California	11,146,000	13,498,000	1.9
Share high technology (%)	6.3	7.9	

Source: State of California (1981), p. 8.

A combination of the product sophistication and rapid growth definition's was used by the State of California for its Commission on Industrial Innovation (Weiss 1985). It is shown in Table 2.3, unfortunately not identified by SIC. It does not encompass all fast-growing sectors, but only those generally acknowledged to be fabricating a product embodying new scientific knowledge. An improvement on the previous version, it uses forecasted growth as its criterion. Yet it shares all the problems just outlined in using growth as a sector selector.

Research and development intensity

High technology industries are most often identified on the basis of R&D expenditures as a percentage of total industry sales. The "R" in R&D consists of two components: basic and applied research. Basic research refers to scientific exploration for the sake of advancing knowledge. Most of this work is undertaken by scientists within universities and private research institutions. Only a small portion of industrial research fits into this category. Applied research, on the other hand, makes up the majority of industrial research, and is loosely defined as the application of scientific and technical principles with the anticipation of economic returns for the effort.

The "D" in R&D refers to the development stage. Here, processes and products identified in the earlier research phase as having market potential are further tested and may eventually become commercial products or processes. Development is usually more costly, risky, and time-consuming than either basic or applied research. One problem with the R&D measure is that it aggregates over quite different mixes of basic and applied research and the more routine advertising and marketing strategies.

Table 2.4 An R&D-based definition of high tech: USA.

SIC	Industry	R&D as percentage of sales
372/376	aircraft & parts/space vehicles & guided missiles	12.2
357	office computing machines	11.6
366	communication equipment	7.4
367	electronic components & assembly	7.4
283	drugs	6.3
383–7	optical, surgical & photographic instruments	6.3
381–2	scientific & measuring instruments	5.3
28	chemicals	3.6
34	fabricated metals	1.2
26	paper & products	0.9
29	petroleum refining	0.8
33	primary metals	0.8
24,25	wood/furniture	0.7
22,23	textiles/apparel	0.4
20	food & kindred products	0.4
	Industry average	3.0

Source: Technical Marketing Associates (1979).

Conceptually, R&D activity comes close to being a good, quantifiable proxy for technologically sophisticated output. However, it is conventionally measured by categorical expenditures rather than by actual measured effort. As a result, R&D expenditure serves as a good selection criterion for industries in the early stages of product development. It is less likely to identify the more mature and stable high tech industries. As Table 2.4 illustrates, petroleum and chemicals fall relatively far down the list with this criterion – chemicals are only slightly above the industry average for R&D. This measure is particularly misleading for industries such as petroleum, with huge sales figures making up the denominator of the research indicator. Because some important industries would be excluded on the basis of this measurement problem, R&D expenditure appears not to provide a good basis for defining high tech.

Occupational structure

Given these problems with alternative criteria, we chose to define high technology industries on the basis of occupational profile. High tech industries are identified as those in which the proportion of engineers, engineering technicians, computer scientists, life scientists, and mathematicians exceeds the manufacturing average. Conceptually, this definition captures the technical capacity of an industry to harness scientific and technical expertise in the development of new products.

The use of an occupational mix criterion to identify high tech industries is superior to alternative indicators in several respects. First, primary occupational categories are standardized across industries. That is, the same distinctions are made among technical, professional, managerial, skilled, and unskilled work. Secondly, occupational definitions are based on standards of competence that are particularly well defined in relation to technical occupations requiring advanced academic training such as engineers and scientists. In addition, most of these professions are certified by national boards, insuring some degree of consistency across occupational categories. Thirdly, three-digit occupational data are collected on a state basis by the federal government, using "job definition" rather than individual self-selection for determining occupational categories. Thus, the data available for this delineation are more comprehensive, precise, standardized and reliable than alternatives such as R&D as a percentage of sales.

In addition to these data features, this indicator comes closest to encompassing the various connotations of high tech. It is closely linked to sophistication of product line and product process. We can assume that large proportions of this type of personnel imply that innovation is proceeding within the sector. Evidence of innovation, in turn, implies the potential for growth – although since such innovation may be of the process as well as of the product variety, such growth may occur in output but not in employment. Yet it does so in a way which permits us to test this link rather than assume it *a priori*.[2]

Defining industry by occupation data

There are two comprehensive sources of national industry occupational data: the 1970 Census-based industry occupation matrix and the 1980 Occupational Employment Statistics (OES) survey-based matrix. The OES survey is a federal–state cooperative effort undertaken on a three-year cycle. States receive the technical specifications for the survey from the Labor Department. After the completion of a full three-year cycle the manufacturing data are forwarded to the Department of Labor, which then compiles the data into the Industry-Occupational Employment matrix.

For a number of good reasons, the OES-based matrix proves a better source of occupational data than the Census. First, it is a survey of employers while the Census asks individuals to select their own occupational category. Secondly, it uses a more precise schedule of occupations based on skill levels. Thirdly, it uses a more comprehensive and rigorous coding system for specifying occupations within an industry. And finally, it uses a much finer disaggregation of industries and of occupations: 378 separate industries and 1,678 occupational categories, against only 201 industries and 377 occupations in the Census.[3]

Using the OES data, we explored two alternative ways of using occupational mix to identify high tech industries at national level. The first used three occupations – engineers, engineering technicians and computer scientists – as a percentage of total industrial employment. In Table 2.5, the right-hand column shows the results: it lists those industries in which the proportion exceeded the national percentage (5.51%) for all manufacturing industries. The second alternative added two extra occupational categories – scientists (including, most importantly, chemists, as well as geologists, physicists, and biological scientists) and mathematicians. Many such scientists, we believe, are engaged in the conceptualization and then the development of new products, for instance, biotechnology and the development of new materials. Though their inclusion did not significantly affect the rank ordering, it did add a few industries to the high tech list (agricultural chemicals, soap, paints, reclaimed rubber, medical and dental supply) while removing two others (fabricated metal products and electrical equipment and supplies). We decided to adopt this alternative, broader definition based on a cutoff point of all sectors exceeding the national average.

The resulting list of 29 three-digit manufacturing sectors is shown in rank order in Table 2.5. Two points should be made about it. First, it is limited to manufacturing sectors; it excludes important service sectors, such as computer software and commercial R&D labs, which undoubtedly deserve the label of high tech. This is regrettable; the fact is

Table 2.5 The 29 high tech sectors and their occupational mix.

Rank	SIC	Title	(a) Engineering/ engineering technicians/ computer scientists	(b) Life & physical scientists	(c) Mathematics	Total a, b & c
			Percentage of total employment			
		Total manufacturing	5.51	0.26	0.05	5.82
1	376	space vehicles & guided missiles	40.90	0.21	0.08	41.19
2	357	office computing machines	26.62	0.05	0.03	26.70
3	381	engineering, laboratory instruments, & scientific instruments	25.67	0.73	0.05	26.45
4	366	communications equipment	21.30	0.26	0.30	21.86
5	383	optical instruments & lenses	18.73	1.03	0.04	19.80
6	286	industrial organic chemicals	14.51	4.85	0.24	19.60
7	372	aircraft & parts	17.95	0.24	0.34	18.53
8	283	drugs	8.86	8.59	0.22	17.67
9	291	petroleum refining	11.76	2.42	0.44	14.62
10	382	measuring & controlling instruments	13.93	0.12	0.09	14.14
11	367	electronic components & assembly	12.72	0.10	0.02	12.84
12	281	industrial inorganic chemicals	9.46	3.14	0.05	12.65
13	282	plastics & synthetic resins	9.38	1.81	0.17	11.36
14	351	engines & turbines	10.16	0.48	0.01	10.65
15	348	ordnance	9.37	0.99	0.06	10.42
16	289	miscellaneous chemicals	6.35	3.70	0.05	10.10
17	386	photographic equipment	8.67	0.80	0.01	9.48
18	362	electrical industrial apparatus	9.24	0.03	0.03	9.30
19	361	electrical transmission equipment	8.55	0.03	0.01	8.59

Table 2.5 The 29 high tech sectors and their occupational mix – *continued*.

Rank	SIC	Title	(a) Engineering/ engineering technicians/ computer scientists	(b) Life & physical scientists	(c) Mathematics	Total a, b & c
			Percentage of total employment			
20	353	construction equipment	8.34	0.05	0.04	8.43
21	285	paints & varnishes	3.22	4.97	0.01	8.20
22	303	reclaimed rubber	5.26	2.27	0.00	7.53
23	356	general industrial machinery	7.21	0.04	0.02	7.27
24	374	railroad equipment	6.58	0.08	0.09	6.75
25	365	radio & TV receiving equipment	6.62	0.06	0.04	6.72
26	287	agricultural chemicals	4.58	1.79	0.11	6.48
27	354	metal working machinery	6.27	0.01	0.00	6.28
28	384	medical & dental supplies	5.42	0.57	0.04	6.03
29	284	soap	3.14	2.71	0.06	5.91

that we could not obtain adequate geographically disaggregated data for such service sectors. But it is mitigated, we think, by the fact that many such service sectors are geographically linked to manufacturing sectors, so that their inclusion would not greatly alter our conclusions about location.

Secondly, the definition is based on three-digit industries, which we term *industrial sectors*. We could not make the definition at a four-digit level because the necessary occupational data do not exist. Nevertheless we have chosen to pursue the subsequent locational analysis at a four-digit level, because we believe that many of our readers will prefer to have this level of detail. This analysis, therefore, is based on 100 components – which we term as *industries* – consisting of all the four-digit industries within the three-digit sectors.[4] It is not possible to say whether all of these would independently have qualified as high tech had the occupational data been available. These 100 industries are presented in Table 2.6.

The disaggregation into 100 industries gives real advantages, as the analysis in subsequent chapters will reveal. The actual gains and losses in

Table 2.6 The 100 high tech industries.

SIC	Industry name
2812	alkalies & chlorine
2813	industrial gases
2816	inorganic pigments
2819	industrial inorganic chemicals, NEC*
2821	plastic materials, synthetic resins
2822	synthetic rubber
2823	cellulosic man-made fibers
2824	synthetic organic fibers, except cellulose
2831	biological products
2833	medical, chemical, botanical products
2834	pharmaceutical preparations
2841	soap, other detergents
2842	special cleaning, polishing preparations
2843	surface active finishing agents
2844	perfumes, cosmetics, toilet preparations
2851	paints, varnishes, lacquers, enamels
2861	gum, wood chemicals
2865	cyclic crudes, intermediates, dyes
2869	industrial organic chemicals, NEC
2873	nitrogenous fertilizers
2874	phosphatic fertilizers
2875	fertilizers, mixing only
2879	pesticides, agricultural chemicals, NEC
2891	adhesives, sealants
2892	explosives
2893	printing ink
2895	carbon black
2899	chemicals, chemical preparations, NEC
2911	petroleum refining
3031	reclaimed rubber
3482	small arms ammunition
3483	ammunition, except small arms, NEC
3484	small arms
3489	ordnance, accessories, NEC
3511	steam, gas, hydraulic turbines
3519	internal combustion engines, NEC
3531	construction machine equipment
3532	mining machinery equipment
3533	oilfield machinery equipment
3534	elevators, moving stairways
3535	conveyors, conveying equipment
3536	hoists, industrial cranes, monorail systems

Table 2.6 The 100 high tech industries – *continued*.

SIC	Industry name
3537	industrial trucks, tractors, trailers, stackers
3541	machine tools, metal cutting types
3542	machine tools, metal forming types
3544	specialty dies, die sets, jigs fixtures, industry molds
3545	machine tool accessories, measuring devices
3546	power driven hand tools
3547	rolling mill machinery equipment
3549	metalworking machinery, NEC
3561	pumps, pumping equipment
3562	ball, roller bearings
3563	air, gas compressors
3564	blowers, exhaust, ventilation fans
3565	industrial patterns
3566	speed changers, industrial high drives, gears
3567	industrial process furnaces, ovens
3568	mechanical power transmission equipment, NEC
3569	general industrial machinery equipment, NEC
3573	electronic computing equipment
3574	calculating accounting machines, except electrical computer equipment
3576	scales, balances, except laboratory
3579	office machinery, NEC
3612	power, distribution special transformers
3613	switch gear, switchboard apparatus
3621	motors, generators
3622	industrial controls
3623	welding apparatus, electric
3624	carbon, graphite products
3629	electrical industrial apparatus, NEC
3651	radio, TV receiving sets, except communication types
3652	phono records, pre-recorded magnetic tape
3661	telephone, telegraph apparatus
3662	radio, TV transmitting, signal, detection equipment
3671	cathode ray tubes, NEC
3674	semiconductors, related devices
3675	electronic capacitors
3676	resistors for electronic applications
3677	resistors, electric apparatus
3678	connectors, electronic applications
3679	electronic components, NEC
3721	aircraft
3724	aircraft engines, parts
3728	aircraft parts, auxiliary equipment, NEC
3743	railroad equipment

Table 2.6 The 100 high tech industries – *continued.*

SIC	Industry name
3761	guided missiles, space vehicles
3764	guided missiles, space vehicles, propulsion units
3769	guided missiles, space vehicles, parts, NEC
3795	tanks, tank components
3811	engineering, laboratory, scientific, research instruments
3822	industrial controls for communications and environmental applications
3823	industrial instruments for measurement and display
3824	fluid meters, counting devices
3825	instruments, measuring, testing, electrical, electrical signals
3829	measuring, controlling devices, NEC
3832	optical instruments, lenses
3841	surgical, medical instruments apparatus
3842	orthopedic, prosthetic, surgical applications
3843	dental equipment, supplies
3861	photographic equipment, supplies

* NEC = not elsewhere classified.

jobs prove much larger than analysis at three-digit level would suggest, and the range of variation in growth rates is also much larger. In other words, there is much greater richness of detail. So, except in Chapter 4, when data problems force us back to work at the level of three-digit sectors, the rest of the analysis in this book will take place at the level of the 100 individual four-digit industries.

Notes

1 Although we are concerned here with manufacturing industries, the same measure can be used for other non-manufacturing industries such as services and the transportation, communications, and utilities industries.
2 Since our original work on this indicator, several others have come to similar conclusions (see Harrington & Vinson 1983, and Riche *et al.* 1983). For a comparison of coverage under different definitions, including ours, see OTA 1983.
3 Further details on the two sources are given in Appendix 1.
4 The 29 sectors vary from those containing single industries (e.g. 381 = 3811, engineering, laboratory, scientific, research instruments) to those with nine (e.g. 356, industrial machinery). It is important to keep in mind that this distinction and all subsequent comparative analysis are in some important ways an artifact of the SIC coding procedures. Thus it is not necessarily true that industries identified at the four-digit level are less significant or more

coherent than others classified at the three-digit level. Nor is it necessarily the case that any two four-digit industries will have more in common than any two three-digit sectors. Sectors and industries are of uneven sizes – a particularly important fact to remember when interpreting absolute job and plant gains. For instance, instruments are classified into four or more three-digit SICs, while industrial machinery is encompassed in only one such SIC. As a result, the three-digit and four-digit instrument industries, taken individually, appear rather small in size in comparison.

3
How high tech grows

Gainers and losers in plants and jobs

HIGH TECH, in the popular view, is the machine that will create the new jobs to replace the ones being destroyed in the sunset industries. Even if high tech eliminates jobs in the industries that incorporate its products, so the argument goes, even larger numbers of jobs will be created in the sectors that make the high tech equipment. Thus, although robots may displace auto workers, even more jobs will be created in the machining plants that make the robots. True, some academic studies have questioned this comfortable view – as for instance those from the Science Policy Research Unit at the University of Sussex in England, which have suggested the reverse: that the job destruction capabilities of high tech are far greater than its job creation potential (Freeman 1982, Freeman et al. 1982). In this chapter, we examine the data for our 100 industries to try to judge whether the optimists or the pessimists have the facts on their side.

Recent high tech job and plant growth

The data on American job growth from 1972 to 1981 show clearly that a significant number of net new jobs were created in high tech industries. More than 1 million more jobs existed in these industries by 1981 (column 4 in Table 3.1). In 1981, these industries accounted for almost 5.5 million jobs. Most striking, and certainly evidence that favors the enthusiasm for high tech generally, is the fact that while high tech industries claimed only 27% of all manufacturing jobs in 1981, they accounted for 87% of the net job growth in manufacturing since 1972.

Table 3.1 Jobs and plants in high tech industries, 1972–81.

SIC	Industry name	Jobs, 1981 (000)	Net new jobs 1972–81 (000)	Plants, 1981
2812	alkalies & chlorine	7.5	−5.8	57
2813	industrial gases	8.8	−0.8	515
2816	inorganic pigments	11.8	−1.0	101
2819	industrial inorganic chemicals, NEC	85.9	22.1	585
2821	plastic materials, synthetic resins	57.7	2.9	469
2822	synthetic rubber	11.2	−0.6	85
2823	cellulosic man-made fibers	15.6	−1.5	25
2824	synthetic organic fibers, except cellulose	62.6	−18.5	60
2831	biological products	21.8	11.7	295
2833	medical, chemical, botanical products	17.4	9.6	188
2834	pharmaceutical preparations	130.5	18.5	641
2841	soap, other detergents	36.0	4.5	676
2842	special cleaning, polishing preparations	24.2	−0.9	774
2843	surface active finishing agents	7.9	1.0	188
2844	perfumes, cosmetics, toilet preparations	54.1	5.9	553
2851	paints, varnishes, lacquers, enamels	60.1	−5.8	1,400
2861	gum, wood chemicals	4.7	−1.2	99
2865	cyclic crudes, intermediates, dyes	30.8	2.6	191

Table 3.1 Jobs and plants in high tech industries, 1972–81 – *continued*.

SIC	Industry name	Jobs, 1981 (000)	Net new jobs 1972–81 (000)	Plants, 1981
2869	industrial organic chemicals, NEC	112.0	9.6	577
2873	nitrogenous fertilizers	10.7	1.3	150
2874	phosphatic fertilizers	15.6	0.7	124
2875	fertilizers, mixing only	12.0	0.6	523
2879	pesticides, agricultural chemicals, NEC	17.2	5.0	323
2891	adhesives, sealants	17.0	2.1	603
2892	explosives	12.0	−6.6	105
2893	printing ink	9.9	0.3	457
2895	carbon black	2.3	−0.6	28
2899	chemicals, chemical preparations, NEC	35.4	−1.7	1,172
2911	petroleum refining	108.8	8.0	444
3031	reclaimed rubber	0.7	−0.2	23
3482	small arms ammunition	9.5	−4.4	54
3483	ammunition, except small arms, NEC	21.8	−33.1	65
3484	small arms	20.5	4.4	110
3489	ordnance, accessories, NEC	27.6	3.0	58
3511	steam, gas, hydraulic turbines	36.1	−10.1	83
3519	internal combustion engines, NEC	89.4	19.5	199
3531	construction machine equipment	145.9	12.1	781
3532	mining machinery equipment	28.2	6.9	349
3533	oilfield machinery equipment	95.0	59.1	797
3534	elevators, moving stairways	11.4	−3.6	126
3535	conveyors, conveying equipment	36.6	9.4	583
3536	hoists, industrial cranes, monorail systems	18.7	2.4	255
3537	industrial trucks, tractors, trailers, stackers	25.6	−0.2	423
3541	machine tools, metal cutting types	74.3	21.8	1,039
3542	machine tools, metal forming types	24.0	−0.1	457

Table 3.1 Jobs and plants in high tech industries, 1972–81 – *continued*.

SIC	Industry name	Jobs, 1981 (000)	Net new jobs 1972–81 (000)	Plants, 1981
3544	specialty dies, die sets, jigs fixtures, industry molds	124.2	26.4	7,035
3545	machine tool accessories, measuring devices	62.2	15.5	1,356
3546	power driven hand tools	26.0	2.9	165
3547	rolling mill machinery equipment	6.0	−4.4	70
3549	metalworking machinery, NEC	25.0	11.4	476
3561	pumps, pumping equipment	72.0	16.5	572
3562	ball, roller bearings	53.3	2.4	166
3563	air, gas compressors	32.7	9.8	215
3564	blowers, exhaust, ventilation fans	30.1	6.6	432
3565	industrial patterns	9.2	0.7	801
3566	speed changers, industrial high drives, gears	25.7	3.2	279
3567	industrial process furnaces, ovens	16.7	3.1	308
3568	mechanical power transmission equipment, NEC	30.9	3.2	237
3569	general industrial machinery equipment, NEC	63.2	26.2	1,433
3573	electronic computing equipment	320.7	175.9	1,207
3574	calculating accounting machines, except electrical	15.5	−7.0	50
3576	scales, balances, except laboratory	6.6	−0.1	110
3579	office machinery, NEC	44.6	10.1	173
3612	power, distribution special transformers	45.9	−0.9	275
3613	switch gear, switchboard apparatus	70.9	1.7	581
3621	motors, generators	93.4	3.1	431
3622	industrial controls	67.2	16.1	774
3623	welding apparatus, electric	19.3	3.8	163
3624	carbon, graphite products	13.0	1.7	78
3629	electrical industrial apparatus, NEC	16.8	3.4	215
3651	radio, TV receiving sets, except communication types	60.6	−25.9	439

Table 3.1 Jobs and plants in high tech industries, 1972–81 – *continued*.

SIC	Industry name	Jobs, 1981 (000)	Net new jobs 1972–81 (000)	Plants, 1981
3652	phono records, pre-recorded magnetic tape	17.8	−2.5	471
3661	telephone, telegraph apparatus	147.4	13.0	263
3662	radio, TV transmitting, signal, detection equipment	426.9	107.7	1,927
3671	cathode ray tubes, NEC	36.2	24.8	100
3674	semiconductors, related devices	169.5	71.9	701
3675	electronic capacitors	28.4	0.8	118
3676	resistors for electronic applications	21.7	1.2	90
3677	resistors, electric apparatus	22.5	−1.7	317
3678	connectors, electronic applications	35.3	17.2	143
3679	electronic components, NEC	190.0	89.5	2,737
3721	aircraft	301.1	69.5	153
3724	aircraft engines, parts	140.0	35.3	281
3728	aircraft parts, auxiliary equipment, NEC	140.3	38.1	776
3743	railroad equipment	48.6	−2.2	193
3761	guided missiles, space vehicles	106.5	−11.9	32
3764	guided missiles, space vehicles, propulsion units	26.7	5.9	26
3769	guided missiles, space vehicles, parts, NEC	19.0	1.9	46
3795	tanks, tank components	14.2	8.3	25
3811	engineering, laboratory, scientific, research instruments	43.5	7.0	826
3822	industrial controls for regulators, resistors, communications and environmental applications	32.6	1.9	230
3823	industrial instruments for measurement and display	53.6	18.0	503
3824	fluid meters, counting devices	15.2	6.4	119
3825	instruments, measuring, testing, electrical, electrical signals	94.8	40.1	686
3829	measuring, controlling devices, NEC	30.6	6.0	572
3832	optical instruments, lenses	43.2	24.4	507

How high tech grows

Table 3.1 Jobs and plants in high tech industries, 1972–81 – *continued*.

SIC	Industry name	Jobs, 1981 (000)	Net new jobs 1972–81 (000)	Plants, 1981
3841	surgical, medical instruments apparatus	54.6	20.1	733
3842	orthopedic, prosthetic, surgical applications	64.9	21.0	1,130
3843	dental equipment, supplies	17.4	5.0	447
3861	photographic equipment, supplies	114.2	18.2	694
	All high tech	5,475.2	1,080.7	49,597
	All manufacturing	20,264.0	1,235.3	4,586,510

Note: Tables 3.1–3.5, based on published data, differ in detail from tables in the remainder of the book.
Source: US Department of Commerce, *Annual survey of manufactures, 1981*; Census of Manufacturers, 1977, vol. 1.

In absolute terms, and keeping in mind the caution just registered regarding categorization, the industries which support the largest workforces are those listed in Table 3.2. They are an interesting group. They are not clustered in a few three-digit sectors, but are found rather broadly across the SIC codings. They are led by radio and TV signalling, transmitting and detection equipment, which accounted for 427,000 jobs in 1981 and is highly defense-oriented (58% dependent on Department of Defense demand in 1982). Computers ranked second, with 321,000 jobs in 1981. Other large sectors in order of rank are aircraft, miscellaneous electronic components, semiconductors, telecommunication equipment, and construction machinery.

Industries with the greatest number of plants show distinctly different rankings. Specialty tool and dye makers top the list, with more than 7,000 establishments. Miscellaneous electronic components ranks second with more than 2,700 plants. Others with more than 1,000 establishments are the radio and TV communications equipment (which leads the job rankings), miscellaneous industrial machinery, paints and varnishes, machine tools, computers, miscellaneous chemicals, orthopedic and surgical instruments, and metal cutting machine tools. The disparity in rankings among the two groups is of course accounted for by differences in average plant size. In other words, firms producing in the latter group operate on average smaller plants, and small businesses are much more common.

Table 3.2 Top 15 industries ranked by employment level and plant incidence, 1981.

Rank	SIC	Industry name	Employment levels (000)
1	3662	radio, TV transmitting, signal, detection equipment	426.9
2	3573	electronic computing equipment	320.7
3	3721	aircraft	301.1
4	3679	electronic components, NEC	190.0
5	3674	semiconductors, related devices	169.5
6	3661	telephone, telegraph apparatus	147.4
7	3531	construction machine equipment	145.9
8	3728	aircraft parts, auxiliary equipment, NEC	140.3
9	3724	aircraft engines, parts	140.0
10	2834	pharmaceutical preparations	130.5
11	3544	specialty dyes, die sets, jigs fixtures, industry molds	124.2
12	3861	photographic equipment, supplies	114.2
13	2869	industrial organic chemicals, NEC	112.0
14	2911	petroleum refining	108.8
15	3761	guided missiles, space vehicles	106.5

Rank	SIC	Industry name	Number of plants
1	3544	specialty dyes, die sets, jigs fixtures, industry molds	7,035
2	3679	electronic components, NEC	2,737
3	3662	radio, TV transmitting, signal, detection equipment	1,927
4	3569	general industrial machinery equipment, NEC	1,433
5	2851	paints, varnishes, lacquers, enamels	1,400
6	3545	machine tool accessories, measuring devices	1,356
7	3573	electronic computing equipment	1,207
8	2899	chemicals, chemical preparations, NEC	1,172
9	3842	orthopedic, prosthetic, surgical applications	1,130
10	3541	machine tools, metal cutting types	1,039
11	3811	engineering laboratory, scientific, research instruments	826
12	3565	industrial patterns	801
13	3533	oilfield machinery, equipment	797
14	3531	construction machine equipment	781
15	3728	aircraft parts, auxiliary equipment, NEC	776

But of greatest interest to policy makers and planners are net jobs created in each sector. Here, the rankings produce a list much more consonant with expectations (see Table 3.3). The computing industry added 176,000 jobs from 1972 to 1981, topping the list. It was followed by the radio and TV communications industry (108,000), miscellaneous electronic components (90,000), semiconductors (72,000), aircraft (70,000), oilfield machinery (59,000) and measuring instruments (40,000).

How high tech grows

Table 3.3 Top 30 industries ranked by absolute job gains, 1972–81.

Rank	SIC	Industry name	Net job gain (000)
1	3573	electronic computing equipment	175.9
2	3662	radio, TV transmitting, signal, detection equipment	107.7
3	3679	electronic components, NEC	89.5
4	3674	semiconductors, related devices	71.9
5	3721	aircraft	69.5
6	3533	oilfield machinery, equipment	59.1
7	3825	instruments, measuring, testing, electrical, electrical signals	40.1
8	3728	aircraft parts, auxiliary equipment, NEC	38.1
9	3724	aircraft engines, parts	35.3
10	3544	specialty dies, die sets, jigs fixtures, industry molds	26.4
11	3569	general industrial machinery equipment, NEC	26.2
12	3671	cathode ray tubes, NEC	24.8
13	3832	optical instruments, lenses	24.4
14	2819	industrial inorganic chemicals, NEC	22.1
15	3541	machine tools, metal cutting types	21.8
16	3842	orthopedic, prosthetic, surgical appliances	21.0
17	3841	surgical, medical instruments apparatus	20.1
18	3519	internal combustion engines, NEC	19.5
19	2834	pharmaceutical preparations	18.5
20	3861	photographic equipment, supplies	18.2
21	3823	industrial instruments for measurement and display	18.0
22	3678	connectors, electronic applications	17.2
23	3561	pumps, pumping equipment	16.5
24	3622	industrial controls	16.1
25	3545	machine tool accessories, measuring devices	15.5
26	3661	telephone, telegraph apparatus	13.0
27	3531	construction machine equipment	12.1
28	2831	biological products	11.7
29	3549	metalworking machinery, NEC	11.4
30	3579	office machinery, NEC	10.1

Source: US Department of Commerce, *Annual survey of manufacturers, 1981*; Census of Manufacturers, 1977, vol. 1.

These seven industries total 615,000 jobs, more than 56% of the total net high tech job growth and 50% of all manufacturing employment growth during this period. With the exception of oilfield machinery (which was clearly related to the energy crisis), these industries are all conventionally high tech.[1] The 30 top-growing high tech industries (Table 3.3) accounted for 99% of all net high tech employment growth and 87% of all manufacturing job growth.

However, the gains posted by the apparently high tech sectors are not by any means synonymous with high tech job creation. For instance, the combined net job totals of six prominent high tech sectors – business machines (SIC 357), electronics (367), aircraft (372), communications equipment (366), engineering instruments (381), and measurement devices (382) – still accounted for only 63% of total net new job creation. Various chemical and industrial machinery sectors were also contributors to the observed job gains of the decade.

Absolute net job growth is, however, only one measure of development potential. An even more sensitive indicator is the percentage change in employment. The job totals for 1977 and 1981 are shown in Table 3.4, along with percentage changes over two periods: 1972–7 and 1977–81. On this basis, a yet different configuration of leading high tech growth industries emerges, and their ranks differ depending upon which interval is chosen as a point of reference. Table 3.5 lists the top proportional job gainers in descending order for each period. Energy-related industries are relatively more important in the earlier period, and defense-related sectors in the latter.

In the earlier period, 1972–7, products associated with the energy crisis – such as fluid meters used in industrial energy conservation projects, mining and oilfield machinery – are prominent, in addition to tanks, various biotech products, miscellaneous metalworking machinery, and optical instruments. In the more recent period, energy-oriented sectors have been largely displaced by industries related to the defense build-up that began at the end of the 1970s: guided missiles, space-related aircraft and ordnance industries, computers, semiconductors, and other electronic industries, and scientific instruments. Electronics-related sectors seem to have increased their job growth performance over the decade. Only oilfield machinery and optical instruments ranked in the top ten in both periods, and only four of the sectors that ranked in the top 15 in the earlier period scored similarly in the latter four years.

Yet of all the insights one can glean from just looking at the numbers, perhaps the most striking is the large number of sectors which actually experienced net job decline. Of our 100 sectors, 28 posted net job losses over the 1972–81 period; for the earlier mid-1970s, 33 of them exhibited negative job growth. Some industries lost jobs in the tens of thousands over the decade, most notably large arms ammunition (34,800), consumer radio and TV (25,900) and synthetic fibers (15,700). In some cases, percentage job losses exceeded 20% in the four years 1977–81 alone: alkalies and chlorine (−36%), reclaimed rubber (−22%), rolling mill machinery (−24%), and phonograph records (−23%). The business cycle presumably does not account for the enormity of these reversals; foreign competition, representing the operation of the product cycle, is clearly the explanation in some of these industries.

Table 3.4 Job growth rates for high tech industries, 1972–81.

SIC	Industry name	Jobs, 1972 (000)	Jobs, 1977 (000)	Jobs, 1981 (000)	Percent change 1972–7	Percent change 1977–81
2812	alkalies & chlorine	13.3	11.8	7.5	−11	−36
2813	industrial gases	9.6	7.5	8.8	−22	17
2816	inorganic pigments	12.8	11.9	11.8	−7	−1
2819	industrial inorganic chemicals, NEC	63.8	78.8	85.9	24	9
2821	plastic materials, synthetic resins	54.8	57.2	57.7	4	1
2822	synthetic rubber	11.8	10.0	11.2	−15	12
2823	cellulosic man-made fibers	17.1	16.0	15.6	−6	−2
2824	synthetic organic fibers, except cellulose	78.2	74.0	62.6	−5	−15
2831	biological products	10.1	15.7	21.8	55	39
2833	medical, chemical, botanical products	7.8	14.4	17.4	85	21
2834	pharmaceutical preparations	112.0	126.4	130.5	13	3
2841	soap, other detergents	31.5	32.1	36.0	2	12
2842	special cleaning, polishing preparations	25.1	22.1	24.2	−12	10
2843	surface active finishing agents	6.9	6.6	7.9	−4	20
2844	perfumes, cosmetics, toilet preparations	48.2	50.9	54.1	6	6
2851	paints, varnishes, lacquers, enamels	65.9	61.4	60.1	−7	−2
2861	gum, wood chemicals	5.9	4.8	4.7	−19	−2
2865	cyclic crudes, intermediates, dyes	28.2	35.7	30.8	27	−14
2869	industrial organic chemicals, NEC	102.4	112.3	112.0	10	0
2873	nitrogenous fertilizers	9.4	12.1	10.7	29	−14
2874	phosphatic fertilizers	14.9	14.4	15.6	−3	8
2875	fertilizers, mixing only	11.4	12.4	12.0	9	−3
2879	pesticides, agricultural chemicals, NEC	12.2	15.0	17.2	23	15

Table 3.4 Job growth rates for high tech industries, 1972–81 – *continued*.

SIC	Industry name	Jobs, 1972 (000)	Jobs, 1977 (000)	Jobs, 1981 (000)	Percent change 1972–7	Percent change 1977–81
2891	adhesives, sealants	14.9	16.7	17.0	12	2
2892	explosives	18.6	12.1	12.0	−35	−1
2893	printing ink	9.6	10.1	9.9	5	−2
2895	carbon black	2.9	2.5	2.3	−14	−8
2899	chemicals, chemical preparations, NEC	37.1	35.3	35.4	−5	0
2911	petroleum refining	100.8	102.5	108.8	2	6
3031	reclaimed rubber	0.9	0.9	0.7	0	−22
3482	small arms ammunition	13.9	10.3	9.5	−26	−8
3483	ammunition, except small arms, NEC	54.9	18.9	21.8	−66	15
3484	small arms	16.1	17.5	20.5	9	17
3489	ordnance, accessories, NEC	24.6	23.6	27.6	−4	17
3511	steam, gas, hydraulic turbines	46.2	40.8	36.1	−12	−12
3519	internal combustion engines, NEC	69.9	88.8	89.4	27	1
3531	construction machine equipment	133.8	155.3	145.9	16	−3
3532	mining machinery equipment	21.3	31.4	28.2	47	−10
3533	oilfield machinery equipment	35.9	58.6	95.0	63	62
3534	elevators, moving stairways	15.0	10.2	11.4	−32	12
3535	conveyors, conveying equipment	27.2	32.9	36.6	21	11
3536	hoists, industrial cranes, monorail systems	16.3	15.8	18.7	−3	18
3537	industrial trucks, tractors, trailers, stackers	25.8	28.8	25.6	12	−11
3541	machine tools, metal cutting types	52.5	59.5	74.3	13	25
3542	machine tools, metal forming types	24.1	23.7	24.0	−2	1
3544	specialty dies, die sets, jigs fixtures, industry molds	97.8	105.6	124.2	8	18
3545	machine tool accessories, measuring devices	46.7	54.1	62.2	16	15

3546	power driven hand tools	23.1	27.7	26.0	20	-6
3547	rolling mill machinery equipment	10.4	7.9	6.0	-24	-24
3549	metalworking machinery, NEC	13.6	19.4	25.0	43	29
3561	pumps, pumping equipment	55.5	63.0	72.0	14	14
3562	ball, roller bearings	50.9	50.6	53.3	-1	5
3563	air, gas compressors	22.9	32.0	32.7	40	2
3564	blowers, exhaust, ventilation fans	23.5	28.0	30.1	19	8
3565	industrial patterns	8.5	9.3	9.2	9	-1
3566	speed changers, industrial high drives, gears	22.5	25.3	25.7	12	2
3567	industrial process furnaces, ovens	13.6	15.2	16.7	12	10
3568	mechanical power transmission equipment, NEC	27.7	32.5	30.9	17	-5
3569	general industrial machinery equipment, NEC	37.0	57.5	63.2	55	10
3573	electronic computing equipment	144.8	192.7	320.7	33	66
3574	calculating accounting machines, except electrical computer equipment	22.5	17.1	15.5	-24	-9
3576	scales, balances, except laboratory	6.7	7.1	6.6	6	-7
3579	office machinery, NEC	34.5	42.4	44.6	23	5
3612	power, distribution special transformers	46.8	43.3	45.9	-7	6
3613	switch gear, switchboard apparatus	69.2	72.0	70.9	4	-2
3621	motors, generators	90.3	96.9	93.4	7	-4
3622	industrial controls	51.1	55.4	67.2	8	21
3623	welding apparatus, electric	15.5	17.5	19.3	13	10
3624	carbon, graphite products	11.3	12.1	13.0	7	7
3629	electrical industrial apparatus, NEC	20.2	16.5	16.8	-18	2
3651	radio, TV receiving sets, except communication types	86.5	74.6	60.6	-14	-19
3652	phono records, pre-recorded magnetic tape	20.3	23.1	17.8	14	-23
3661	telephone, telegraph apparatus	134.4	124.4	147.4	-7	18
3662	radio, TV transmitting, signal, detection equipment	319.2	334.1	426.9	5	28

Table 3.4 Job growth rates for high tech industries, 1972–81 – continued.

SIC	Industry name	Jobs, 1972 (000)	Jobs, 1977 (000)	Jobs, 1981 (000)	Percent change 1972–77	Percent change 1977–81
3671	cathode ray tubes, NEC	11.4	36.7	36.2	222	−1
3674	semiconductors, related devices	97.6	114.0	169.5	17	49
3675	electronic capacitors	27.6	28.9	28.4	5	−2
3676	resistors for electronic applications	20.5	21.3	21.7	4	2
3677	resistors, electric apparatus	24.2	20.7	22.5	−14	9
3678	connectors, electronic applications	18.1	26.0	35.3	44	36
3679	electronic components, NEC	100.5	125.9	190.0	25	51
3721	aircraft	231.8	222.7	301.1	−4	35
3724	aircraft engines, parts	104.7	106.1	140.0	1	32
3728	aircraft parts, auxiliary equipment, NEC	102.2	102.0	140.3	0	38
3743	railroad equipment	50.8	56.3	48.6	11	−14
3761	guided missiles, space vehicles	118.4	94.0	106.5	−21	13
3764	guided missiles, space vehicles, propulsion units	20.8	18.6	26.7	−11	44
3769	guided missiles, space vehicles, parts, NEC	20.9	7.2	19.0	−66	164
3795	tanks, tank components	5.9	12.4	14.2	110	15
3811	engineering, laboratory, scientific, research instruments	36.5	42.3	43.5	16	3
3822	industrial controls regulators, resistors, communications and environmental applications	30.7	39.0	32.6	27	−16
3823	industrial instruments for measurement and display	35.6	46.5	53.6	31	15
3824	fluid meters, counting devices	8.8	15.9	15.2	81	−4
3825	instruments, measuring, testing electrical, electrical signals	54.7	66.5	94.8	22	43

3829	measuring, controlling devices, NEC	24.6	32.3	30.6	31	−5
3832	optical instruments, lenses	18.8	30.0	43.2	60	44
3841	surgical, medical instruments apparatus	34.5	43.2	54.6	25	26
3842	orthopedic, prosthetic, surgical applications	43.9	53.9	64.9	23	20
3843	dental equipment, supplies	12.4	16.3	17.4	31	7
3861	photographic equipment, supplies	96.0	111.7	114.2	16	2
	All high tech	4,394.5	4,747.4	5,475.2	8	15
	All manufacturing	19,028.7	19,590.1	20,264.0	3	3

Source: US Department of Commerce, *Annual survey of manufacturers, 1981*; Census of Manufacturers, 1977, vol. 1.

Table 3.5 Industries ranked by percentage job gains, 1972–7 and 1977–81.

1972–7

Rank	SIC	Industry name	Percent job change
1	3671	cathode ray tubes, NEC	222
2	3795	tanks, tank components	110
3	2833	medical, chemical, botanical products	85
4	3824	fluid meters, counting devices	81
5	3533	oilfield machinery equipment	63
6	3832	optical instruments, lenses	60
7	2831	biological products	55
8	3532	mining machinery equipment	47
9	3678	connectors, electronic applications	44
10	3549	metalworking machinery, NEC	43
11	3563	air, gas compressors	40
12	3573	electronic computing equipment	33
13	3823	industrial instruments for measurement and display	31
13	3829	measuring, controlling devices, NEC	31
13	3843	dental equipment, supplies	31

1977–81

Rank	SIC	Industry name	Percent job change
1	3769	guided missiles, space vehicles, parts, NEC	164
2	3573	electronic computing equipment	66
3	3533	oilfield machinery equipment	62
4	3679	electronic components, NEC	51
5	3674	semiconductors, related devices	49
6	3764	guided missiles, space vehicles, propulsion units	44
6	3832	optical instruments, lenses	44
8	3825	instruments, measuring, testing, electrical, electrical signals	43
9	3728	aircraft parts, auxiliary equipment, NEC	38
10	3678	connectors, electronic applications	36
11	3721	aircraft	35
12	3724	aircraft engines, parts	32
13	3549	metalworking machinery, NEC	29
14	3662	radio, TV transmitting, signal, detection equipment	28
15	3841	surgical, medical instruments apparatus	26

Source: US Department of Commerce, *Annual survey of manufactures, 1981*; Census of Manufacturers, 1977, vol. 1.

Indeed, relatively poor job growth rates are a common rather than aberrant phenomenon among these high tech industries. Forty-three industries grew at the same rate or more slowly than manufacturing in general during the period 1977–81, compared to 38 during the earlier period. These sectors accounted for more than 1.7 million jobs, or about

32% of all high tech jobs in 1981. In addition, a decline of approximately 100,000 jobs were recorded in individual sectors over the period 1977–81, so that a gross 828,000 jobs had to be created to generate the net job gain of 728,000. This dislocation must be taken into account when evaluating the performance of high tech industries.

The unavoidable conclusion is that though most high tech industries are rapidly growing or are at least net job generators, some are very definitely not. Further, it needs underlining that the analysis in this chapter refers to the whole period 1972–81: four years longer than the analyses in the chapters that follow, and years in which high tech industries experienced relatively high growth rates compared with the mid-1970s. If anything, therefore, this chapter has given a favorable view of the job-creation potential of high tech.

Note

1   Data on plant distribution was not available from the Annual Survey for 1981, so that we could not easily compare net plant additions for the whole period. Data on plants is available from the County Business Patterns, but there are fairly large discrepancies in the reporting and aggregation procedures of the two data sources. An earlier tabulation (see Glasmeier *et al.* 1983) showed that over the period 1972–7 many fewer sectors experienced plant loss than job loss, reflecting both the rationalization of production at existing plants and the economy-wide phenomenon of a boom in small-plant openings.

4
Why high tech grows

The product–profit cycle

HIGH TECH INDUSTRIES are not a uniform or homogeneous group. Though most grow, some do not. The obvious question is whether there is any general explanation for these differences. In this chapter we draw on theory to present a simple explanatory model. The theory, the product–profit cycle, suggests that firms in any one industry will go through successive stages of development, from youth through maturity to old age. When we classify our high tech sectors according to this model, we find that they range from dynamic, innovative high growth performers to mature, even declining, sectors with poor job-creation prospects. We also encounter a group of sectors which defy classification according to this evolutionary scheme; their upswings and downswings appear to correspond to defense spending cycles. And, finally, we discover one anomaly: that the most high tech industries, in the sense that they have the highest proportions of scientific and technical workers, are not necessarily the fastest growing.

Stages in high tech evolution

Product–profit cycle theory suggests that as industries – and the firms in these industries – evolve from one stage to another, so they will

Why high tech grows 41

experience striking variations in growth rate, profitability, occupational mix, market power, and locational patterns. These differences occur because in the different stages they will develop distinct, and generally sequential, business strategies.[1] Below, we outline these changes as they occur in a four-stage model. Then, we seek to distinguish the unique features of military-related sectors.

Before we do so, however, one point needs to be stressed. The original product cycle, as developed by Vernon (1966), was of relatively short duration (6–8 years) and referred to a product in a fairly narrow sense, as for instance a new kind of machine tool, or a particular model of an automobile. Much subsequent literature on product cycles, however, has applied the theory in a wider sense and on a longer timescale, to refer to the development of whole industries as a result of major technological innovations (for instance, the automobile industry or the aircraft industry). These latter applications draw on an older theory of innovation, as developed first by Kondratieff (1935) and then by Schumpeter (1939), which suggests that bursts of fundamental innovations occur at approximately 57-year intervals, producing so-called long waves of economic development (Mandel 1980, van Duijn 1983). Though we do not pretend here to comment on long-wave theory, throughout this chapter – and in the rest of this book – when we refer to product–profit cycles we do so in this wider sense.

Innovation

In its earliest stage of development, a new sector is distinguished by its preoccupation with the design and commercialization of a new product. The sector consists of a number of firms, predominantly new start-ups, which pursue the adaption of an innovation; it is this uniqueness of product and the phenomenon of emerging new firms which results in the sector's recognition, and its assignment of a new SIC number.[2] Extraordinarily high profits may accompany commercialization, and these are generally plowed back immediately into the firm.

The major efforts of firms in this phase are bent toward perfecting the product, which is often customized, produced in small batches, and continually redesigned to client specification. Thus large proportions of the workforce consist of engineers, technicians, and other specialists. At the same time, commercialization results in rapid growth in the workforce, including management, sales and production workers. Firms who initially secure patents may enjoy a brief period of market power, enabling them to price the product relatively high, but entry by new start-ups and more established firms will soon place pressure on prices and erode the early leaders' edge.

The preoccupation with product design and commercialization generally reinforces the geographical concentration of the sector's activity within initial centers. These clusters may be built around older centers of innovation, or they may be rather accidental in their location, due to an inventor's or firm founder's hometown preferences. As new firms enter, they will gravitate toward the sector's geographical core, because it is there that information sources about the day-to-day changes in the market are most easily tapped and where supplier services tailored to the sector, such as specialized machine tool and venture capital firms, are to be found. The rapid agglomeration of firms, suppliers and other related activities may create intense pressures on land costs and environmental quality in these places, crowding out other unrelated businesses and creating enclaves of upper income households.

Market penetration

Once the product design is fairly well worked out and the result standardized, the major preoccupation of the firm shifts towards mass production and market penetration. At this stage, good management and sales personnel become essential and account for a relatively larger share of new employees than at other stages. Growth rates in output and employment will continue to rise during this period, although at a more modest rate than previously, just as profit margins will converge towards "normal" rates of return. Market power will continue to erode, unless a well-financed marketing effort, generally by a conglomerate backer, manages to create name recognition and a sales/service network that squeezes smaller competitors out.

During this phase, the sector's organizational build-up will continue to add employment in original centers. However, a net dispersal of employment will begin overall, as the more standardized portions of the production process are spun off to cheaper (especially lower land and labor cost) locations. This dispersal is apt to be confined to sites within the same region at this stage – largely peripheral land parcels on the fringes of the same metropolitan areas or rural areas within the same region. The market penetration effort will also disperse sales personnel across regions, although this will probably not show up as shifts in manufacturing employment because these workers are employed by separate retail organizations. The beginnings of this dispersal will leave the original centers even more specialized than previously, as they become new headquarters enclaves with smaller proportions of production workers.

Market saturation

Once the major markets have been reached, firms in an industry settle in to vie for market shares. Two strategies exist in this phase. The first (and the only one under competitive conditions) is to produce your standardized product more cheaply than your rivals, increasing your market share by underpricing them in the market. Cost-cutting and/or productivity gains are the only means of maintaining acceptable profit levels. Firms operating with this strategy are preoccupied with cutting costs, which may include pursuing economies of scale, purchasing cheaper inputs, cutting transportation costs, adapting process innovations to raise labor productivity, and tightening labor discipline.

The second variation occurs when a small number of firms achieve market power and are able to collude to restrict output, stabilize market shares, and secure oligopolistic profits. Instead of competing on the basis of price, firms in this configuration will be preoccupied with market management, product differentiation, observation of rivals, and labor peace. They will employ relatively larger shares of management and administrative personnel than their price-competing counterparts and be more apt to tolerate unions. Effective oligopolies are most apt to be formed in this stage, although industries with low start-up costs and few economies of scale may never go this route. Whichever regime characterizes the sector in this phase, it will experience only modest output growth and will exhibit even less employment growth.

In the market saturation phase, decentralization of production will accelerate, especially at the interregional (and international) level. Cheaper labor costs, better business climates and proximity to markets will pull the more standardized portions of the production process away from innovating centers. In the case of oligopolistic sectors, this process will be more polarized. On the one hand, the larger share of the workforce preoccupied with market management, observation of rivals, and product differentiation will anchor the sector around its original core. On the other hand, the relatively larger size of firms and ample corporation resources, expanded by oligopolistic returns, may accelerate the redeployment of production in outlying areas. In either case, this process will tend to draw upon quite different resources at each location, reinforcing the differentiation of communities within the interregional division of labor.

Rationalization

In the fourth stage, intensified competition drives the price of the product below an acceptable rate of return. This may occur because imports penetrate the regional or national market, because new technologies displace the function of the product, or because a cheaper, substitute commodity serves end users better. Firms in the industry will selectively close down capacity in response, a process often referred to as "rationalization" (Massey & Meegan 1982). Some plants will continue to be operated, even at a loss, to cover fixed costs and to hedge against possible revival of the market. But the firms' overall strategy will be to diversify out of the product line or to meet importers' lower production costs by overseas relocation. In this stage, employment will decline rather precipitously, as layoffs accompany plant closings.

In this era, two tendencies act in a contradictory fashion upon market concentration. If the difficulties facing the industry are viewed as temporary, the larger, more diversified firms may endure while smaller firms fail, thereby increasing concentration of market power in the industry. On the other hand, if the reversal of fortunes seems permanent, the larger firms are often the first to rationalize, leaving small firms to struggle on in an increasingly competitive environment.

During this phase, the industry will tend to reconcentrate production in a few locations. Reconcentration in newer, outlying sites typifies those industries who dispersed in search of lower-cost production sites. The older, more out of date facilities in the historical and higher cost center are closed first, resulting in a new concentration of employment in regions to which the industry more recently dispersed. Examples are the steel industry's recent exodus from Pittsburgh, recentering around Chicago, and the textile industry's postwar shift from New England to the Southeast, both taking place in tandem with sustained net employment decline.

Reconcentration may also reflect the abandonment of newer, outlying facilities which were mainly market penetration efforts. This alternative pattern is exemplified by the auto industry which recently closed many of its postwar market-oriented assembly plants, recentering production around Detroit. Because of the heavy incidence of plant closings in this stage, reconcentration of either one type or the other is bound to prevail in contrast to the dispersal of previous stages.

Military-related sectors
a special case

One group of high tech industries can be distinguished by their unusual mission — serving the US military. These sectors exhibit a unique corporate structure, distinctive production strategies and unique locational tendencies. Since they cannot easily be understood within the framework of the evolutionary analysis just described, they deserve separate treatment.[3]

First of all, the nature of demand is almost entirely different from other manufacturing sectors. Military goods are sold in the main to one buyer, the federal government, who is thus a monopsonist. But unlike private sector monopsonists, this customer does not use its market power to depress the price of the product (and therefore the profits of sellers) in the market transaction. Many authors have stressed the importance of performance and time as dominant selection criteria for the Department of Defense; demand is price inelastic and may be zero during wartime. Cost insensitivity has been institutionalized into the cost-plus contract, where overruns in excess of 100% are common in larger weapons and equipment procurement programs (Baldwin 1967, Gansler 1980, Melman 1974).

On the supply side, these industries are highly oligopolistic. A small number of corporations do the bulk of Defense Department business (Adams 1981). The top 100 companies account for 70% of total business, the top 25 for 50% and the top 5 for 20% (Gansler 1980). Concentration ratios at the sectoral level are high and have been increasing over time (Baldwin 1967). At the level of individual products, firms generally compete with only one or two others for the initial contract for a new system, and after that exercise a virtual monopoly. Subcontracting is considerably more competitive, but in this arena, too, concentration has been growing rapidly, in part because the larger prime contractors have increasingly chosen to produce "in-house" than to buy from others. Often subcontractors have survived by occupying market niches and becoming monopolists of certain highly specific parts (Gansler 1980).

The nature of the military product and its production process is also quite distinct from commercial manufacturing. The units of output of assembled weapons, equipment and transport systems, are often (and increasingly) few in number and highly sophisticated in nature. The product embodies constant innovation, requiring large numbers of scientific and technical personnel engaged in design, prototype building and testing. Skilled labor for assembly (e.g. machinists) is also a large cost component. In contrast, materials account for a relatively small proportion of total costs. Often immense in size (e.g. submarines, military transport planes, missiles), these products tend to be produced in very large-scale facilities. Batch rather than assembly line production predominates, so that defense-oriented plants are more like huge craft shops than standardized mass production factories.

For these reasons, we would expect military equipment, weapons and transport to be distinctively "high tech" in nature. Cold war rivalries have resulted in the institutionalization of continual innovation in weaponry and delivery systems. The goal of military R&D is to make the product of the opposition ineffective, and the interaction of international competition of this sort, particularly between the Soviet Union and the USA, has been to create a dynamic wherein existing products are rapidly rendered obsolete. Product innovation, then, rather than process innovation, dominates industrial research and development activities in these sectors, creating what Malecki (1983) calls technology push in place of demand pull as the growth trajectory. These sectors are thus more or less permanently locked into the initial stages of their product cycles, except that should an entire system be displaced by another (battleships by submarines and aircraft), their demise is sudden and final.

Because of this constant product innovation and because the products themselves are highly sophisticated, large numbers of scientists and technicians are employed by these military suppliers. Engineers alone accounted for relatively larger percentages of the workforce in aircraft (11%), communications equipment (15%), guided missiles (31%), and ordnance (5%) than in manufacturing as a whole (3%) (DeGrasse 1983). Three major military-oriented categories – aircraft and missiles, electrical equipment and communications, and machinery – accounted for one in three R&D scientists and engineers in a 1967 National Science Foundation study (Dumas 1982). Defense-generated employment was skewed in 1970 toward the professional and technical categories in comparison with all US employment (Dempsey & Schmude 1971). Several studies have attempted to estimate the percentages of scientific and engineering jobs attributable to direct and indirect impacts of defense spending. One found that defense spending accounted for 48% of aeronautical engineers, 23% of physicists, 21% of electrical engineers, 19% of all mathematicians and 16% of industrial engineers, and that 38% of physicists, 22% of electrical engineers, 20% of technical engineers, 20% of mechanical engineers, and 20% of metallurgical engineers were employed either as civilians directly by the Defense Department or by defense-related industries (Rutzick 1970). Given the heavy reliance of military producers on this type of scientific personnel, it is logical to anticipate that these sectors are apt to figure prominently among high tech industries, however defined.

For a number of reasons, we expected industries with heavy reliance on military markets to be relatively more concentrated geographically than other sectors. Proximity to military bases, and the need for extensive, inexpensive and sometimes secret sites for large-scale weapons system assembly and testing, will encourage concentration. At the same time, the forces normally operating to disperse maturing industries – market penetration and low labor costs – are of diminished importance in sectors less sensitive to cost differentials and more preoccupied with performance.

Grouping high tech sectors by stage

Conceptually, we would like to be able to locate different industries within the profit cycle by direct evidence on profitability. But, for a number of reasons, this is impossible. Profit data are recorded at the corporate rather than establishment level; this gives disproportionate weight to large diversified corporations, so that major differences in market conditions across product lines are cancelled out. Small, single-product firms whose profitability might be a better gauge of sectoral circumstances are underrepresented.[4] Alternatively we could identify the progress of an industry through the cycle by qualitative data on business strategy. However, collection of such data would be laborious and would involve difficult judgments. For some phases, such as market penetration, satisfactory indicators would be hard to come by.

So, perforce, we adopt a second-best measure: we measure the path of employment growth over time. The theory is that during the innovative, superprofits period, employment will grow at dramatic rates. During the era of market penetration, manufacturing employment will grow, but at more modest rates. Once the market is saturated and cost containment dominates, employment growth will level off and begin to decline, even when output is still growing. If oligopoly prevails in this stage, the slowdown in job growth will be postponed. Once heightened competition from substitutes or imports begins consistently to squeeze profit rates, job growth rates will be negative. If oligopolistic market power has retarded the productivity growth (and job decline) process, the adjustment during this rationalizing period may be quite abrupt, causing extensive worker dislocation.

In other words, employment growth rates are hypothesized to approximate a normal curve over time. One way of measuring product–profit cycles, therefore, is to find the inflexion points in this curve.[5] In order to do this, we needed employment data over a reasonably long period of time, approximating to the entire post-World War II era. Such data are available only at the three-digit level, for the good reason that many of our individual high tech industries did not even exist in the late 1940s. So, for this analysis, we had to work with our 29 sectors rather than with the 100 separate industries (Table 4.1 below).

For each of these sectors, we classed as *innovative* all sectors with postwar compound annual growth rates in excess of 3% whose growth had not peaked as of 1981. *Market penetrating*, modest growth sectors were classified as those with growth rates between 1.0% and 2.9% per year, whose job levels had reached peaks in either 1977 or 1981. *Mature, market saturation* sectors were identified as those whose compound annual growth rates were less than 2.0% for the entire postwar period and below 0.59% since 1967, with job levels peaking before 1977. *Profit-squeezed, rationalizing* sectors posted negative job growth rates in the postwar period and peaked in 1954 or previously. Sectors could only be allocated according to this scheme if they exhibited fairly simple sinusoidal-shaped growth curves. Those with more than one prominent peak had to be classified separately as "volatile."

Using compound annual growth rates[6] and evidence on the peaking of employment levels,[7] we grouped our 29 high tech industries into the five groups shown in Table 4.1. For each group, the table shows the compound annual growth rate for 1947–81 and also for 1967–81, a period most observers regard as characterized by profound structural change. These values are calculated on a simple straight-line basis between the two end-dates, a procedure that could be criticized statistically because of possible distorted values at these dates. So the table also gives an alternative calculation for 1947–81 based on ordinary least squares regression, a statistical device often used to overcome the problem. The column labelled 'R^2' shows the correlation between the two calculations; as can be seen, this is sometimes high but quite often low. The final column of Table 4.1 shows the year in which employment peaked.

Group A, the innovators, consists of six sectors with compound annual job growth rates in excess of 3% and employment levels which have not yet peaked (i.e. the highest levels were for the most recent year, 1981). This group consists of those predominantly capital goods sectors commonly thought of as high tech: electronics, computers, communications equipment, medical technology, and two groups of instruments. Postwar job growth is charted for each member of this group in Figure 4.1. All except optical instruments show relatively consistent growth paths, reflected in high R^2 in the table. Communications equipment and electronic components both exhibit a rather severe post-Vietnam recession, but returned to high growth rates in the late 1970s.[8]

Group B, the market penetrators, consists of seven sectors whose compound growth rates fall in the range 1% to 3% per year and whose

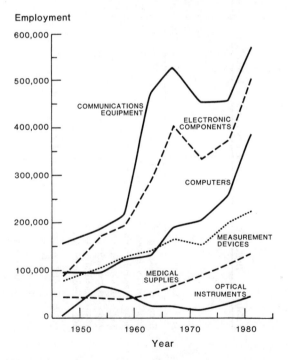

Figure 4.1 Employment change, innovating industries, 1947–81.

Table 4.1 High tech sectors classified by profit cycle location, 1947–81.

		Two-point formula 1947–81	Compound annual growth rate Regression estimate 1947–81	R^2	Two-point formula 1967–81	Peak years
SIC	Name					
Group A Rapid growth sectors: innovation						
383	optical instruments & lenses*	5.7	−0.1†	0.00	3.9	1981
367	electronic components & assembly*	5.2	4.8	0.90	1.6	1981
357	office computing machines*	4.4	5.9	0.85	5.2	1981
366	communications equipment*	3.9	3.9	0.78	0.6	1981
384	medical & dental supplies	3.5	5.3	0.86	5.2	1981
382	measuring & controlling instruments	3.1	3.1	0.95	2.3	1981
Group B Modest growth sectors: market penetration						
386	photographic equipment	2.4	2.9	0.92	1.3	1981
283	drugs	2.2	2.4	0.92	2.6	1981
353	construction equipment	1.8	2.2	0.88	2.0	1981
284	soap	1.5	1.6	0.95	1.5	1981
356	general industrial machinery*	1.4	1.6	0.89	1.3	1981
286	industrial organic chemicals	1.3	1.5	0.96	0.9	1977
351	engines & turbines	1.1	1.4	0.82	1.6	1977

Group C Volatile sectors: defense dependency					
376	space vehicles and guided missiles*	7.6	2.4†	0.22	-4.1	1967,1981
348	ordnance*	4.4	3.3†	0.25	-5.7	1967,1981
372	aircraft & parts*	3.1	0.2†	0.01	-1.6	1958,1967,1981
381	engineering, laboratory and scientific instruments*	2.2	0.5	0.03	-0.4	1958,1967,1981
354	metalworking machinery*	1.2	1.0	0.54	0.1	1981,1967,1954
289	miscellaneous chemicals*	0.9	0.7	0.29	-1.4	1967,1954
362	electrical industrial apparatus*	0.6	0.7	0.54	0.1	1981,1967,1947
Group D Mature sectors: market saturation					
365	radio & TV receiving equipment	1.4	1.5	0.33	-3.6	1967
282	plastics & synthetic resins	1.3	1.8	0.82	-0.1	1972
361	electrical transmission equipment	1.1	1.2	0.79	0.1	1981,1972
281	industrial inorganic chemicals*	1.1	0.3	0.07	0.3	1954
287	agricultural chemicals	0.7	0.4	0.35	-0.2	1967
285	paints & varnishes	0.4	0.5	0.61	-0.4	1972
Group E Declining sectors: rationalization					
291	petroleum refining	-0.9	-1.3	0.76	0.1	1954
374	railroad equipment	-1.8	-1.1	0.22	-1.1	1947
303	reclaimed rubber	-3.1	-3.8	0.76	-6.1	1954

*These sectors contain one or more four-digit sectors identified by the Bureau of Industrial Economics (Henry 1983) as part of the "Defense Industrial Base."

†Due to lack of reliable data for the 1947 to 1954 period, and to the use of a minimum estimate, these compound growth rates are over-estimated.

employment levels have peaked only recently, if at all (either 1977 or 1981). These sectors cover a range of chemical machinery and instruments sectors, largely capital goods but including two (drugs, photographic equipment) with significant consumer sales. Within this group, satisfactory correspondence exists between rankings generated with all three growth rates. Growth paths over the postwar period show some sensitivity to the business cycle, especially after 1967, but as a whole these sectors sustained modest growth rates through the late 1970s (Fig. 4.2).

Group C consists of seven sectors which for lack of a better label we have classified as volatile. These sectors display two, sometimes three, definite peaks in employment in the postwar period. In no case were 1977 or 1981 employment levels significantly above the 1967 peak. They defy

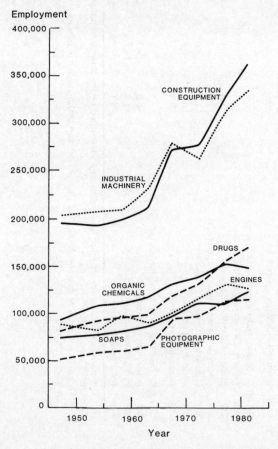

Figure 4.2 Employment change, market penetrating industries, 1947–81.

Why high tech grows

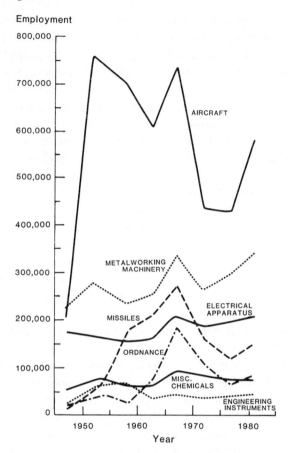

Figure 4.3 Employment change, volatile industries, 1947–81.

characterization by their location along a bell-shaped employment growth path, as the erratic growth patterns displayed in Figure 4.3 and the very low R^2s indicate.

Two interpretations of this group's erratic performance are possible. One is that because its production is biased toward producer goods, its markets are simply more highly sensitive to swings in the business cycle than are other sectors. Indeed, the peaks do seem to correspond to boom Census years. An alternative explanation is that these sectors are much more dependent upon military spending than others in the high tech set. This is more satisfactory for explaining the prominent peaking around 1981, which cannot be considered a boom year; it is also consistent with the strong peaking around 1967, one of the final years of the steamed up Vietnam economy.

A preliminary test of this latter hypothesis can be carried out by comparing this group of high tech sectors with the Bureau of Industrial Economics list of defense-related industries (Henry 1983). Using a conservative definition of defense spending,[9] the BIE includes the sectors starred in Table 4.1 in its "Defense Industrial Base," although the proportions of defense demand vary widely. All sectors falling into volatile Group C encompass at least one defense-dependent four-digit sector. Some are highly defense-dependent. Shares of 1982 output directly and indirectly attributable to the Department of Defense range from 91% (ordnance), 68% (missiles), and 41% (miscellaneous chemicals, mainly explosives) to 28% (engineering instruments), 6% (metalworking machinery) and 8% (electrical apparatus).

This test suggests a compromise interpretation of Group C in which the first four of these high tech sectors – missiles, ordnance, engineering instruments, and aircraft – exhibit volatile growth paths because of defense dependency, while the other three industrial equipment suppliers are particularly sensitive to the business cycle, reflecting the accelerator principle. It is interesting that only the defense-dependent sectors among this group have posted overall growth rates that would qualify them for Group A.

Defense-related sectors are also heavily represented in Group A. Five out of six of the fast-growing sectors are on the defense industrial base list. One, radio and TV communications, is quite heavily dependent on defense, with 58% of output going to the Department of Defense; another, optical instruments, is moderately so with 28%, while the others display ratios below 20%. It should be noted, however, that the defense share of output for these premier high tech sectors is growing rapidly under the Reagan military build-up. For instance, in 1979 only 3.6% of computer output went to DOD, compared to 12.7% forecast for 1987, reflecting an increase of 141% in defense demand (Henry 1983).

Another way of underscoring the prominent role of defense spending in the performance of high tech sectors is to note that the defense-based sectors are almost entirely clustered in Groups A and C, the rapid growth and volatile sectors. Of 16 modest growth, mature and declining sectors, only two – industrial machinery and inorganic chemicals – qualify on the defense industrial base list. And looking at it in another way, all but one of the nine sectors which achieved compound growth rates in excess of 3% per year in the postwar period were members of the defense industrial base. None of the six sectors which experienced sustained negative job growth over the whole period 1947–81 were on the same

Why high tech grows 55

list. Defense spending seems also to be an important factor in distinguishing rapid from modest growth sectors. While five out of six of the Group A sectors formed part of the defense industrial base, only one out of seven in Group B did.

The fourth group consists of six mature sectors subsisting in relatively saturated markets, often with serious import challenges. Their growth rates have been less than 2% over the postwar period, less than 0.5% since 1967, and they reached peak employment in a Census year before 1977.[10] This group displays humped job growth curves; in all cases these sectors did fairly well in the boom period up through 1967 and have failed to create jobs as rapidly since (see Fig. 4.4). This explains the relatively low R^2 for the estimated coefficients. Four of the six actually posted negative job growth rates after 1967. This group is skewed toward industrial commodities rather than equipment, with the exception of electrical transmitters, the consumer-oriented TV and radio receivers and similar portions of paints and varnishes. Of this nationalizing group, only inorganic chemicals has an appreciable defense-related market.

A final set of three sectors posted negative job growth rates for the postwar period (Fig. 4.5). Their employment levels each peaked in or before 1954. These are sectors for which demand has declined due to new substitutes (trucking for railroads, synthetics for rubber) or in which automation has reduced employment levels dramatically (petroleum refining). Since 1967, job decline has slowed, due to the energy crisis.

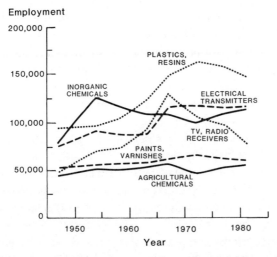

Figure 4.4 Employment change, market saturated mature industries, 1947–81.

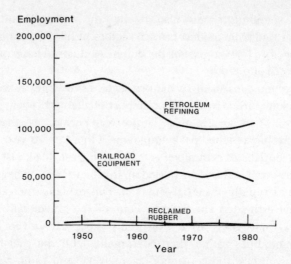

Figure 4.5 Employment change, rationalizing (declining) industries, 1947–81.

Expanded demand for increased domestic oil actually increased employment slightly. The revival of the coal industry produced a derived demand for railroad equipment, slowing that sector's decline.

These latter two groups are particularly interesting because they show that sectors with a sophisticated product, and/or one produced with a sophisticated production process (as in the case of oil refining), may not be job creators. Most of these sectors had relatively high proportions of engineers and scientists working for them as of the late 1970s, yet some had been declining in employment for years. We conclude from this that high tech character, as defined in the previous chapter, does not guarantee the job expansion potential of an industry. Since we are interested in job generation, we could alternatively select just those three-digit sectors which are growing rapidly. Yet currently this would mean including some, like apparel, aluminum, and coal processing, that are not high tech in any other sense – not in R&D, nor shares of professional/technical workers, nor perceived sophistication of product.

Nor do we find a statistically significant relationship[11] between sectors with high shares of professional/technical workers and relatively higher job growth rates. We experimented with a number of possible forms of a regression equation in which the compound sectoral growth rates were hypothesized to be a function of the proportions of engineers, engineering technicians, computer scientists and other scientists and mathematicians. Although there was a relationship, in that the slope generated was always positive, it was statistically a very weak one: in no case did the R^2 exceed 0.05. This exercise confirmed our hypothesis that high proportions of certain technical occupations do not guarantee a sector's prospects as an employment generator. There are sectors (such as oil) which have a very high proportion of scientists and technicians, but which have low growth rates in the period concerned; equally, there are sectors where the reverse obtained. This needs emphasizing, because it has obvious implications for the development of an industrial policy at both national and local levels.

Notes

1. This discussion draws heavily upon Markusen (1985a). For an application to international trade, see Vernon (1966).
2. Of course, the designation process is imperfect and often lags behind the reality. An example is that computers were not given a separate SIC number until 1967, about a decade after their debut as a product quite different from other business machines. Sectors often encompass several products, each with its own dynamic; for empirical purposes, it is essential to begin by reading the SIC handbook. Furthermore, almost all new sectors have at least some older firms competing along with the new. Computers illustrate this mix as well, where IBM competes alongside the Apple.
3. For an extended discussion of these features, see Markusen (1985a).
4. For an extended discussion of these problems, see Markusen (1985a), Ch. 7.
5. In the jargon of calculus, the first and second derivatives.
6. There is no *a priori* reason for choosing average (straight-line) annual growth rates over compound ones. The former assumes that growth arises from a stable base, i.e. new employment does not affect the rate at which the economic base generates additional employment. The compound growth rate is preferable when describing a process whereby growth arises from an expanding base and one wishes to assume that new employment increases the rate at which employment in the next period is generated. In reality, we are trying to model a process which we have hypothesized to be shaped like a bell-shaped curve, so that neither is really perfect. Since we expect on balance that the high tech industries will be concentrated on the growth end of the curve, we chose the compound rate.
7. Employment series were constructed from the Census of Manufacturers. Since the Census frequently redefines sectors (and this is especially true of emergent and high tech industries) but does not publish historical series, it was necessary in a large number of cases to reconstruct them. The conversion tables at the end of each Census volume show the percentages of employment in previous Census year constituent sectors. We built our series by assuming that these proportions held further back in time, so that each time a sector was divided or aggregated these coefficients were used to estimate historical levels. In four cases (SICs 348, 372, 376, and 383) it proved impossible to estimate 1947 levels fully. In these cases, minimum levels were used. The regression estimates are thus likely to be biased upward for these four sectors.

8 Readers should note that the graphs show absolute employment levels, so that larger sectors, which appear higher up on the figures, produced more net new jobs without growing faster in terms of percentage change.
9 The definition used in the BIE's analysis includes only Department of Defense expenditures and not NASA, Department of Energy and other military-related outlays; it is unclear whether arms exports are fully accounted for. The underestimation of defense demand is clearly reflected in the fact that shares for missiles and ordnance, which are almost totally sold to defense markets, are not higher. See Shutt (1984).
10 SIC 361, electrical transmitters, actually peaked in 1981, but its employment level has not changed substantially since 1972.
11 That is, a linear, monotonic relationship. There may well be some more complex type of relationship, but this is not what interests us here.

5

How high tech organizes

Competition, monopoly, and the profit cycle

IN THIS CHAPTER and the next, we continue to test the validity of product–profit cycle theory. That theory, as set out in Chapter 4, hypothesized that market power in innovative high tech industries should be quite highly concentrated either because of patents or the novelty of the product, or both. This advantage of pioneer firms is subsequently eroded by heightened competition in the struggle to penetrate the market. Following that, in an era of market saturation, market power is again consolidated, albeit fashioned with different tools.

Market power has traditionally been gauged by economists with the device of the concentration ratio, a measure which simply sums the percentage of sales or value added which the top 4, 8, 20 or 50 firms account for in a particular industry (Mueller 1970, Scherer 1980). While the ratio probably underestimates true market power (Markusen 1985a), it does offer a minimal estimate of control. Therefore, we use it as a measure in the analysis that follows.

For most of our 100 four-digit industries, we were able to capture the change in market power from 1963 to 1977. This relatively short period was insufficient for drawing definitive conclusions regarding the

Table 5.1 Concentration ratios for high tech industries, 1977.

SIC	Name	Four-firm	Eight-firm	Twenty-firm	Change in eight-firm 1963–77
2812	alkalies & chlorine	66	87	99+	−1
2813	industrial gases	65	84	93	−2
2816	inorganic pigments	54	78	94	−6
2819	industrial inorganic chemicals, NEC	33	49	74	−3*
2821	plastic materials, synthetic resins	22	37	60	−4*
2822	synthetic rubber	60	83	99	+3
2823	cellulosic man-made fibers	D	100	100	0
2824	synthetic organic fibers, except cellulose	78	90	99	−9
2831	biological products	32	49	74	−8
2833	medical, chemical, botanical products	65	78	89	−1
2834	pharmaceutical preparations	24	43	73	+5
2841	soap, other detergents	59	71	82	−9
2842	special cleaning, polishing preparations	41	56	67	+2*
2843	surface active finishing agents	32	46	69	−4
2844	perfumes, cosmetics, toilet preparations	40	56	74	+4
2851	paints, varnishes, lacquers, enamels	24	36	52	+2
2861	gum, wood chemicals	59	80	92	+9
2865	cyclic crudes, intermediates, dyes	42	60	79	−11
2869	industrial organic chemicals, NEC	38	55	73	−8
2873	nitrogenous fertilizers	34	54	82	+1*
2874	phosphatic fertilizers	35	57	92	+10*
2875	fertilizers, mixing only	21	37	56	+6
2879	pesticides, agricultural chemicals, NEC	44	64	78	+7*
2891	adhesives, sealants	24	35	54	+4*
2892	explosives	64	79	96	−7
2893	printing ink	43	57	74	−6
2895	carbon black	70	100	100	0
2899	chemicals, chemical preparations, NEC	15	25	40	−5
2911	petroleum refining	30	53	81	−3
3031	reclaimed rubber	74	D	D	n.a.

Table 5.1 Concentration ratios for high tech industries, 1977 – *continued*

SIC	Name	Four-firm	Eight-firm	Twenty-firm	Change in eight-firm 1963–77
3482	small arms ammunition	86	97	99	+2*
3483	ammunition, except small arms, NEC	52	71	93	+9*
3484	small arms	58	78	95	+4*
3489	ordnance, accessories, NEC	48	72	91	+4*
3511	steam, gas, hydraulic turbines	86	97	88	−1
3519	internal combustion engines, NEC	49	70	88	+5
3531	construction machine equipment	47	59	75	+6
3532	mining machinery equipment	37	50	70	−3
3533	oilfield machinery equipment	30	45	66	+8
3534	elevators, moving stairways	52	68	82	−5
3535	conveyors, conveying equipment	19	30	46	−9
3536	hoists, industrial cranes, monorail systems	16	30	57	−23
3537	industrial trucks, tractors, trailers, stackers	45	61	75	−3
3541	machine tools, metal cutting types	22	35	56	+3
3542	machine tools, metal forming types	18	32	55	−7
3544	specialty dies, die sets, jigs fixtures, industrial molds	8	10	14	+1
3545	machine tool accessories, measuring devices	20	31	45	+2
3546	power driven hand tools	50	70	95	n.a.
3547	rolling mill machinery equipment	62	77	93	−11*
3549	metalworking machinery, NEC	15	24	43	−3*
3561	pumps, pumping equipment	17	29	52	+2*
3562	ball, roller bearings	56	71	88	−5
3563	air, gas compressors	45	64	86	−8
3564	blowers, exhaust, ventilation fans	17	28	49	−15
3565	industrial patterns	9	14	23	0
3566	speed changers, industrial high drives, gears	29	42	64	+1*
3567	industrial process furnaces, ovens	26	39	58	−3
3568	mechanical power transmission equipment, NEC	26	44	69	−14*

Table 5.1 Concentration ratios for high tech industries, 1977 – *continued*.

SIC	Name	Four-firm	Eight-firm	Twenty-firm	Change in eight-firm 1963–77
3569	general industrial machinery equipment, NEC	10	16	27	−3†
3573	electronic computing equipment	44	55	71	−28†
3574	calculating accounting machines, except electrical computer equipment	59	83	98	−7†
3576	scales, balances, except laboratory	50	66	85	−3
3579	office machinery, NEC	60	76	88	+5
3612	power, distribution special transformers	56	70	85	−9
3613	switch gear, switchboard apparatus	51	65	78	0
3621	motors, generators	42	55	72	−4
3622	industrial controls	42	54	68	−15
3623	welding apparatus, electric	47	65	80	+11
3624	carbon, graphite products	80	88	97	−4
3629	electrical industrial apparatus, NEC	28	43	66	−8
3651	radio, TV receiving sets, except communication types	51	65	81	+3
3652	phono records, pre-recorded magnetic tape	48	62	75	−13
3661	telephone, telegraph apparatus	D	89	94	−7
3662	radio, TV transmitting, signal, detection equipment	20	33	57	−12
3671	cathode ray tubes, NEC	58	78	95	n.a.
3674	semiconductors, related devices	42	62	79	−3
3675	electronic capacitors	47	62	86	+1*
3676	resistors for electronic apparatus	38	63	86	−2*
3677	resistors, electric apparatus	20	30	46	+6*
3678	connectors, electronic applications	45	65	82	−6*
3679	electronic components, NEC	29	36	45	−5*
3721	aircraft	59	81	99	−2
3724	aircraft engines, parts	74	86	93	−1*
3728	aircraft parts, auxiliary equipment, NEC	45	56	73	+5*
3743	railroad equipment	51	65	86	−5*
3761	guided missiles, space vehicles	64	94	100	+6*

Table 5.1 Concentration ratios for high tech industries, 1977 – *continued*.

SIC	Name	Four-firm	Eight-firm	Twenty-firm	Change in eight-firm 1963–77
3764	guided missiles, space vehicles, propulsion units	69	93	100	+1*
3769	guided missiles, space vehicles, parts, NEC	76	86	97	+1*
3795	tanks, tank components	87	97	100	−2*
3811	engineering laboratory, scientific, research instruments	25	37	56	−3
3822	industrial controls for regulators, resistors, communications and environmental applications	59	81	92	+6
3823	industrial instruments for measurement and display	32	46	64	−7*
3824	fluid meters, counting devices	43	67	87	−10
3825	instruments, measuring, testing, electrical, electrical signals	33	43	61	−4
3829	measuring, controlling devices, NEC	25	35	52	−2*
3832	optical instruments, lenses	30	43	63	−10
3841	surgical, medical instruments apparatus	32	48	67	−10
3842	orthopedic, prosthetic, surgical applications	38	49	62	−9
3843	dental equipment, supplies	33	46	69	−4
3861	photographic equipment, supplies	72	86	90	+10

† Change in eight-firm ratio reflects 1967–77.
* Change in eight-firm ratio reflects 1972–7.
D = withheld to avoid disclosing operations of individual companies.

performance of individual industries over time. However, it is possible to compare the industries on a cross-sectional basis, to see if market power varies by degree of maturity at one point in time. By doing so, we can also draw some inferences about relative market power for sectoral groupings.

Overall, market power does exist in high tech sectors (Table 5.1). Of the four-digit industries, 38 exhibited 1977 four-firm concentration ratios of 50% or over. We have classified these as "highly oligopolized" and listed

Table 5.2 Highly oligopolized high tech industries.

SIC	Industry	Four-firm
2823	cellulosic man-made fibers	90+*
3661	telephone, telegraph apparatus	90+*
3795	tanks, tank components	87
3482	small arms ammunition	86
3511	steam, gas, hydraulic turbines	86
3624	carbon, graphite products	80
2824	synthetic organic fibers, except cellulose	78
3769	guided missiles, space vehicles, parts, NEC	76
3031	reclaimed rubber	74
3724	aircraft engines, parts	74
3861	photographic equipment, supplies	72
2895	carbon black	70
3764	guided missiles, space vehicles, propulsion units	69
2812	alkalies & chlorine	66
2813	industrial gases	65
2833	medical, chemical, botanical products	65
2892	explosives	64
3761	guided missiles, space vehicles	64
3547	rolling mill machinery equipment	62
2822	synthetic rubber	60
3579	office machinery, NEC	60
2841	soap, other detergents	59
2861	gum, wood chemicals	59
3574	calculating accounting machines, except electrical computer equipment	59
3822	auto, controls regulators, resistors, communications, environmental applications	59
3721	aircraft	59
3484	small arms	58
3671	cathode ray tubes, NEC	58
3562	ball, roller bearings	56
3612	power, distribution special transformers	56
2816	inorganic pigments	54
3483	ammunition, except small arms, NEC	52
3534	elevators, moving stairways	52
3613	switch gear, switchboard apparatus	51
3651	radio, TV receiving sets, except communication types	51
3743	railroad equipment	51
3576	scales, balances, except laboratory	50
3546	power driven hand tools	50

*Two sectors were so highly concentrated that four-firm ratios were suppressed:

2823	cellulosic man-made fibers	90+
	1977 four-firm withheld to avoid disclosing operations of individual companies.	
	1977 eight-firm = 100; 1972 four-firm = 96.	
3661	telephone, telegraph apparatus	90+
	1972 & 1977 four-firm withheld to avoid disclosing operations of individual companies (footnote D in Table 5.1).	
	1977 eight-firm = 89; 1970 four-firm = 94.	

How high tech organizes

them in Table 5.2, ranked in descending order. Another 26 industries had eight-firm ratios of 50% or over in 1977, a group we have classified as "moderately oligopolized" and arrayed in Table 5.3. Together, these two sets of industries comprise almost two-thirds of our total. The remaining industries we have classified as "competitive" even though large firms may play a prominent role; they are listed in Table 5.4.

Profit cycle stages and market power

Generally, the more mature sectors did indeed exhibit a greater tendency toward higher concentration ratios. When grouped according to their stages as identified by long-term growth paths, those industries in rationalizing or market saturation sectors had disproportionately high ratios. Indeed, all the declining industries, and three-quarters of market saturated industries, had eight-firm ratios above 50%.

Table 5.3 Moderately oligopolized high tech industries.

SIC	Industry	Eight-firm
3489	ordnance, accessories, NEC	72
3519	internal combustion engines, NEC	70
3824	fluid meters, counting devices	67
3623	welding apparatus, electric	65
3678	connectors, electronic applications	65
2879	pesticides, agricultural chemicals, NEC	64
3563	air, gas compressors	64
3676	resistors for electronic applications	63
3652	phonograph records, pre-recorded magnetic tape	62
3674	semiconductors, related devices	62
3675	electronic capacitors	62
3537	industrial trucks, tractors, trailers, stackers	61
2865	cyclic crudes, intermediates, dyes	60
3531	construction machine equipment	59
2874	phosphatic fertilizers	57
2893	printing ink	57
2842	special cleaning, polishing preparations	56
2844	perfumes, cosmetics, toilet preparations	56
3728	aircraft parts, auxiliary equipment, NEC	56
2869	industrial organic chemicals, NEC	55
3573	electronic computing equipment	55
3621	motors, generators	55
2873	nitrogenous fertilizers	54
3622	industrial controls	54
2911	petroleum refining	53
3532	mining machinery equipment	50

Table 5.4 Competitive high tech industries.

SIC	Industry	Eight-firm
2819	industrial inorganic chemicals, NEC	49
2831	biological products	49
3842	orthopedic, prosthetic, surgical applications	49
3841	surgical, medical instruments apparatus	48
2843	surface active finishing agents	46
3823	industrial instruments for measurement and display	46
3843	dental equipment, supplies	46
3533	oilfield machinery equipment	45
3568	mechanical power transmission equipment, NEC	44
2834	pharmaceutical preparations	43
3629	electrical industrial apparatus, NEC	43
3825	instruments, measuring, testing, electrical, electrical signals	43
3832	optical instruments, lenses	43
3566	speed changers, industrial high drives, gears	42
3567	industrial process furnaces, ovens	39
2821	plastics materials, synthetic resins	37
2875	fertilizers, mixing only	37
3811	engineering, laboratory, scientific, research instruments	37
2851	paints, varnishes, lacquers, enamels	36
3679	electronic components, NEC	36
2891	adhesives, sealants	35
3541	machine tools, metal cutting types	35
3829	measuring, controlling devices, NEC	35
3662	radio, TV transmitting, signal, detection equipment	33
3542	machine tools, metal forming types	32
3545	machine tool accessories, measuring devices	31
3535	conveyors, conveying equipment	30
3536	hoists, industrial cranes, monorail systems	30
3677	resistors, electric apparatus	30
3561	pumps, pumping equipment	29
3564	blowers, exhaust, ventilation fans	28
2899	chemicals, chemical preparations, NEC	25
3549	metalworking machinery, NEC	24
3569	general industrial machinery equipment, NEC	16
3565	industrial patterns	14
3544	specialty dies, die sets, jigs fixtures, industrial molds	10

The innovative and market penetrating industries possess a more diverse mix of industry structures. In the rapidly growing innovative sectors, highly competitive as well as strongly concentrated industries exist side by side. Within these, computers, office machines and telecommunications are dominated by a small number of firms, while electronics segments are modestly concentrated and the scientific instruments are relatively competitive. However, it is also true that in the case of the latter, the categories themselves obscure a great deal of specialization within the industry. Individual instruments firms may in fact serve market niches which they monopolize, so that the concentration ratio does not accurately reflect true market power.

In comparison, the market penetrating industries do appear to be relatively more competitive than the innovative sectors. This may be interpreted as support for the theory that new firm entry pursues an initial innovative advantage and will engender greater competition.

Within the sectors characterized by volatile long-term growth patterns, those dependent upon military spending show disproportionately high levels of concentration, especially in the aircraft and missiles industries. On the other hand, concentration ratios were relatively low in the slower growing of these sectors, those tagged in Chapter 6 as cyclically sensitive producers' goods industries.

Clearly, some of the differences within profit cycle stage groupings are due to variations in economies of scale, not only in production but also in marketing. But the evidence permits us to conclude that an initial oligopolistic position is likely to be challenged by new entrants in a market-penetrating stage, and that market power will re-emerge in later stages of maturation. On the other hand, we detected a strikingly high degree of market power in some innovative industries, including some electronic components, telecommunications equipment, environmental controls, and office machines. This phenomenon lends credence to arguments that underlying structural tendencies in the economy, namely the emergence of the conglomerate form and the integration of the world market with its requirements to operate on a global scale, are accelerating the pace of concentration in innovative industries and may eventually eliminate the intermediate, more competitive period (Markusen 1985a).

Market power in contemporary fast-growing industries

Corroborating evidence can be sought in the degree to which those industries identified as rapid net job creators in the most recent period have shown a tendency to become more competitive than the relatively sluggish high tech sectors. We did find that net job creating sectors displayed a tendency, over the period 1963-77, to become more competitive, as measured by a fall in their concentration ratios.

ratios. Indeed, all the declining industries, and three-quarters of market saturated industries, had eight-firm ratios above 50%.

Symmetrically, market power appeared to increase disproportionately in net job loss industries.

However, it is dangerous to generalize from either growth experience. Net job growth industries are not universally exhibiting successful firm entry, nor are all net job loss industries experiencing heightened market power. In both cases, approximately one-third of the sectors studied showed the opposite tendency. In other words, wide variations within fast growing and negative job growth groupings caution against assuming that market power will automatically fall with market success or rise with a slowdown in growth of demand.

Market power in military-related sectors

Military-related high tech manufacturing is characterized by substantial degrees of market power. Concentration is particularly acute in the large price-ticket items, such as aircraft, missiles, space vehicles, tanks and ordnance (Table 5.5). Of the 27 industries with the greatest dependence (exceeding 16% of sales) on military markets, only seven show eight-firm ratios below 50%. These latter fall mainly in the instruments and electronics components sectors.

The finding of high degrees of market power in military-related high tech dovetails with the findings of much of the literature on industrial structure among lead firms participating in the military-industrial complex. A small number of corporations do the bulk of Defense Department business (Adams 1981). The top 100 companies account for 70% of total business, the top 25 for 50%, and the top 5 for 20% (Gansler 1980). Concentration ratios are not only high, but have been increasing historically (Baldwin 1967). At the level of individual products, firms generally compete with only one or two others for the initial contract for a new system and after that exercise a virtual monopoly. Subcontracting is considerably more competitive, but in this arena, too, concentration has been growing rapidly, in part because the larger prime contractors have increasingly chosen to produce "in-house" rather than buy from others. Often subcontractors have survived by occupying market niches and becoming monopolists of certain highly specific parts (Gansler 1980).

How high tech organizes 69

Table 5.5 Concentration ratios in defense-related industries.

SIC	Industry name	Percent military*	Eight-firm concentration ratio 1977
3489	ordnance, accessories, NEC	100.0	72
3483	ammunition, except small arms, NEC	99.7	71
3795	tanks, tank components	99.2	97
3484	small arms	66.0	78
3761	guided missiles, space vehicles	65.1	94
3721	aircraft	64.9	81
3724	aircraft engines, parts	63.6	86
3764	guided missiles, space vehicles, propulsion units	63.6	93
3662	radio, TV transmitting, signal, detection equipment	60.5	33
3728	aircraft parts, auxiliary equipment, NEC	56.5	56
3769	guided missiles, space vehicles, parts, NEC	56.5	86
3811	engineering, laboratory, scientific, research instruments	49.2	37
2892	explosives	43.6	79
3822	auto, controls regulators, resistors, communications, environmental applications	34.6	81
3823	industrial instruments for measurement and display	34.6	46
3824	fluid meters, counting devices	34.6	67
3829	measuring, controlling devices, NEC	34.6	35
3675	electronic capacitors	29.1	62
3676	resistors for electronic applications	29.1	63
3677	resistors, electric apparatus	29.1	30
3678	connectors, electronic applications	29.1	65
3679	electronic components, NEC	29.1	36
3537	industrial trucks, tractors, trailers, stackers	26.5	61
3674	semiconductors, related devices	22.5	62
3511	steam, gas, hydraulic turbines	22.1	97
3671	cathode ray tubes, NEC	18.1	78
3832	optical instruments, lenses	16.9	43

* Shutt 1984, data from Department of Defense DEIMS model, for 1983.

Summary

By disaggregating the three-digit sectors chosen in the previous chapter into their four-digit constituents, we have demonstrated the complexity of high tech development performance and industrial structure. Constituent elements of three-digit sectors do not necessarily share the same fortunes. In some cases, they do exhibit relatively similar job growth patterns and market concentration ratios, as in the instruments industries. Yet even here, market power within SIC 382 varies from eight-firm ratios of 35% for measuring and control devices to 81% for environmental controls. In other cases, job losses in one four-digit subsector seem to be compensated for by substitutes in the same sector – for instance, electron tubes are apparently succumbing to competition from semiconductors. In yet other cases, rapid growth in one four-digit SIC is neither complemented by nor counteracted by growth or decline in a sibling.

At the four-digit level, it is even clearer that high tech industries vary dramatically in their ability to generate net job growth. Not only is this variation striking across the set, but individual industries that are high job producers in one era may be the scene of job elimination in the next. Failure of individual high tech industries to grow at the same rate as manufacturing as a whole was quite common among industries in our set. Absolute decline also occurred. Indeed, more than 100,000 jobs disappeared in these industries in the most recent period despite the overall contribution of high tech to job creation.

Market power proves to be considerable in a majority of high tech sectors, challenging the popular image of high tech as a relatively competitive, cut-throat environment. Indeed, the two are not really incompatible. Rather, the control of many markets by a small number of firms simply means that competition may be taking different forms than traditional price rivalry: instead, firms compete through product differentiation, market penetration and advertising. The resources required to compete on this basis appear to be pushing many smaller firms into mergers with larger competitors or into conglomerate partnerships.

6

How high tech clusters

Geographical concentration and dispersal

IF PRODUCT–PROFIT cycle theory is valid, then it should also apply to the ways in which industries – and the firms in them – locate. Innovative high tech industries, still in their youthful developmental stages, should cluster together. Later, as they enter the stage of market penetration, they should begin to disperse. Finally, in their mature stage, they should exhibit a high degree of dispersal across the nation and even across the world. To test this thesis for American high tech, in this chapter we develop measures of spatial concentration for two years in the 1970s for which detailed Census of Manufacturers data were available: 1972 and 1977. We also calculate measures of plant and job redistribution. Thus, while continuing to answer *how?* questions, this chapter begins to turn to the *where?* questions, which provide the focus for the second half of the book.

Some theory
forces for agglomeration and dispersal

Theorists, especially those attempting to generalize about industrial location on a macro scale, have long been drawn to the idea that firms find it advantageous to locate in industrial complexes to a degree not explained by the micro-economic calculus for each individual unit. The explanations center upon the notion of "agglomeration," which refers to the external economies generated for all participants by proximity to each other (as buyers, sellers or competitors), and by proximity to a common resource pool, particularly labor, on which all may draw. Beginning with Marshall (1890), many studies have elaborated upon this theme, most extensively that of Richardson (1973). Yet the concept of agglomeration remains disturbingly residual in nature – a sort of catchall term for spatial concentrations of industry which cannot be explained through any other, or at least distinguishable, variable.

Even before the term agglomeration became formalized, locational theorists were noticing the centralizing tendencies of innovation and its power to promote extraordinarily high rates of urbanization and growth in host areas (Thompson 1962, 1975, Friedmann 1972, Pred 1975). Others were beginning to note that as industries became increasingly dominated by routine production processes, they spread to locations other than the original center, a process Hoover attributed to the lessening in need for highly specialized labor and the increasingly repellant effect of high wages and inflexible working conditions associated with a skilled elite (Hoover 1948). International business analysts have built models along the same lines which predict the specialization of such countries as the USA in commodities which are highly scientific-labor-intensive (Vernon 1966, Hirsch 1967, Wells 1972), while standardized productive activities would be relocated to less developed countries. Focusing on regions within the USA, Thomas (1975, 1981) extended the product cycle model to link dispersal with successive evolutionary stages.

More recently, a number of theorists have linked the emergence of a new configuration of leading edge industries with new centers of agglomeration (Malecki 1980a, b, Markusen 1985a). The centripetal forces driving Silicon Valley, Route 128 around Boston, the twin cities of Minneapolis–St. Paul, Austin and Huntsville are an amalgam of factors on both demand and supply sides, coupled with crucial informational links. While for analytical purposes these can be listed and even ranked in importance, in reality they have been elements in an historical process in which it is hard to isolate any single one as the central formative influence.

First of all, there are the frequently anomalous circumstances that account for the birth of an industry in the first place. At least one researcher argues that a majority of new firms sprout in founders' home towns (Taylor 1975). Several prominent cases of this have been noted in the USA. For instance, Saxenian (1985) shows that Frederick Terman's residence at Stanford was a chief factor in the genesis of Silicon Valley. On the other hand, Feller (1975) shows with the example of light bulbs that there is no necessary relationship between the location of inventive activity and that of successful commercialization.

More often, the success of a single entrepreneur or upstart firm is related to institutional factors which promote one region's advantages over others. The origins of the Route 128 phenomenon in the Boston area are strongly attributed to commercial spin offs of research undertaken at the area's major universities, much of it funded by defense contracts (Dorfman 1983). The same may be said of Stanford's role in engendering Silicon Valley. Sometimes private corporate research labs are the source of new innovations and firm spin offs, again often with defense-related seed money. Examples are the role that Minnesota Mining and Manufacturing (3M) and Honeywell played historically in forming the high tech complex in the Minneapolis–St. Paul area. New government initiatives, such as the space program, have clearly underwritten the emergence of yet other high tech centers, such as Houston, Huntsville (Alabama) and Melbourne–Titusville (Florida). Military bases, as the locations of hardware testing facilities and the final destination for assembled war material, figured prominently in the historical siting of aircraft and related electronics activities. Los Angeles is the most obvious example of this link.

Whatever the initial location, its head start is reinforced by several features of the growth dynamic. First of all, information needs compel would-be competitors to cluster closely around the pioneering firms. Changes in the new product and reorganization of its fabrication and marketing take place on an almost daily basis, so that word of mouth becomes a primary means of data gathering for key decision making. Newer entrants are frequently set up by engineers and other professionals leaving the pioneer firms. Between 1959 and 1979, for instance, Fairchild Semiconductor spawned 50 new high tech companies in Santa Clara County (Saxenian 1985). Knowing the area resources and already having an ear to the ground on new developments, these new aspirants are highly likely to remain nearby. Informational imperatives are shored up as new specialist firms in market research spring up around the high tech core, and by customer beliefs that only firms with a "high tech" address will be "in the know." Software firms interviewed in the early 1980s stated that a Silicon Valley, Palo Alto or San Francisco letterhead was far preferable to one from Oakland (Hall *et al.* 1985).

As clusters of new productive activity sprout, a labor force tailored to the peculiar needs of that sector begins to form. Generally, this is not an indigenous labor pool, but is recruited either directly by rapidly expanding firms or indirectly through the agency of local universities. The vast majority of Silicon Valley engineers, for instance, have been drawn there over the past 20 years by either Stanford and Berkeley campuses or by firms' own recruitment efforts. The larger firms, such as Lockheed, do more of the interregional recruiting and serve as a sort of conduit for engineers and technicians entering the local labor pool. A 1984 interview with Lockheed recruiters indicates that of the 700 or so new engineers recruited a year, at least half come from outside of California, especially from such midwestern engineering schools as Iowa, Illinois, Minnesota, and Purdue (Deitrick 1984). In the earliest days of a new high tech complex, production workers may also be recruited interregionally, especially in occupations such as machinists which are highly skilled. But as standardization and automation progress, this becomes less common.

Members of this transplanted labor force develop an institutional knowledge about the area's firms and a stake in the options available to them regarding mobility among employers. It becomes increasingly difficult to entice this type of worker away from leading centers of innovation, even when cost of living may be lower and pay levels higher at alternative sites. (Workers over the age of 30 may represent an exception.) Thus, over time, the formation of a highly specialized labor force, composed of professionals and technicians themselves preoccupied with innovative activities, reinforces the agglomerative tendencies of these industries. Presumably, management and marketing personnel respond to the same incentives.

Most new high tech firms are small, even personal, ventures. The principal of the firm is likely to be an engineer or technician with a better idea who believes he or she can make a go of it independently and who has the courage to try. This type of entrepreneur often lacks the basic business skills so essential to survival. For this reason, a set of small, specialized business service firms often spring up around new high tech centers, offering marketing, accounting, consulting and financial services to high tech hopefuls. The growth of this supporting sector in the area's economy in turn reinforces its nodal position. Some such services actually become nationally prominent in the design of organizational and financial structures tailored to an emerging industry; an example is the evolution of a unique venture capital market for electronics in Silicon Valley. New start-ups who challenge the industry's leaders remain clustered around the pioneers, not so much for information about product developments but to draw upon the market research firms and the specialists in financing that the pioneers have drawn in their wake.

One final factor in the reinforcement of initial agglomerative tendencies is the willingness of customers to come to the seller in early stages of a product's development. In many cases, important users of high tech products will locate their plants near the seller. A striking example is the clustering of computer manufacturers around the home of semiconductors in Silicon Valley. In turn, computer software firms have gravitated toward centers of innovation in computing. Smaller businesses who use high tech products often send their buyers to the source of innovation for purchasing; for instance, Brazilian professionals and technicians are frequently sent by their companies and government agencies to Palo Alto to purchase needed computer hardware. The location of major high tech users alongside their suppliers, and the willingness of others to travel to high tech centers to purchase, is a powerful factor in the structuring of high tech complexes and in their persistence.

Fundamental to the process we have just described is the fact that new high tech firms are relatively undifferentiated internally. Initially one building serves as headquarters, marketing and sales center, research and design lab, production facility and warehouse. While some of these functions might be less expensively performed elsewhere, the firm has not yet reached the threshold where commercial success, economies of scale, and financial resources permit the separation of production from planning and marketing functions.

The significance of the ability of the modern industrial corporation to distinguish these functions and locate them at separate sites was first elaborated on by Hymer (1972, 1979). The product cycle theorists cited above have also hypothesized that the more routine portions of the production process will successively be spun off to lower-cost branch plant locations, while corporate headquarters with their financial, strategic coordination and research functions remain at the original centers. Markusen (1985a) has argued that the emergence of oligopoly early in an industry's evaluation will retard this process of decentralization, but that the erosion of market power may accelerate it at later stages.

The decentralization of production is not only due to high costs repelling plants from centers of innovation but is also a response to the decentralized locations of buyers and users. While the costs of product transportation are less important to high tech industries than they are to steel or autos, the increased emphasis on just-in-time production, relatively customized products, and prompt servicing encourages large-scale producers to site production facilities in regional centers, economies of scale permitting.

Nevertheless, spatially differentiated cost factors are also significant. Lower labor costs have been a major impetus for the offshore siting of semiconductor assembly plants and, more recently, in the electronics business as witnessed by the Atari case. And it is not only production labor costs that encourage dispersal, but the high cost of professional labor in areas where the housing prices are abnormally inflated from an overly successful local economy. A major demand of high tech executives in California in recent years has been for state housing assistance. Five US semiconductor companies have located their more routine research facilities in Israel, where engineers are willing to work for salaries at about 40% of those in the USA.

Space costs are also a factor. Production facilities are generally more land-extensive than are office functions, and land in places like Silicon Valley is available only at the premium. Indeed, so intense is the competition for industrial land, and so negative the perceived consequences for cost and quality of residential life, that several local governments in Silicon Valley have contemplated (and in some cases adopted) industrial growth controls in the form of moratoriums on new construction, office density limits, and incentives for van pooling and other techniques of abating traffic congestion. The associated costs of space as well as access costs for those portions of the industry shipping products in and out of the region have encouraged the dispersion of production from such centers of innovation.

This model describes the development path of firms which form new centers of innovation. It leaves open the question of organizational structure in places which are not such initial innovation centers, but are rather the recipients of dispersal as firms within these industries mature. As this happens, there develops a whole continuum of spatial-organizational relationships; it includes primary centers of basic research and development populated by entrepreneurial ventures and branches of large corporations, secondary concentrations of technical branch establishments undertaking product-line R&D as well as assembly and production, and tertiary clusters of standardized branch production and assembly facilities. Focusing on the organizational structure of firms at either end of this cycle overlooks the important intermediate stages.

Glasmeier (1986) presents a perspective on such a continuum by arguing that the expansion path of high tech firms in core areas, such as Route 128 and Silicon Valley, has been through the creation of "technical branch plants." These facilities form the economic base of secondary high tech centers such as Austin, Texas; Phoenix, Arizona; and Colorado Springs, Colorado. The work of Gordon and Kimball (1986) on Santa Cruz suggests that it displays similar features, although, because firms here tend to be small, independent and close to Silicon Valley, not identical.

Technical branch plants are organized around product lines and include research and development functions as well as fabrication and assembly. Unlike the "small" entrepreneurial firm hypothesized in the early stage of the product cycle, technical branch plants are relatively large, independent profit centers responsible for the continued development and production of sophisticated technology-rich products. They must successfully compete in their respective markets just as entrepreneurial ventures do, or go out of business. What sets them apart from the small entrepreneurial venture is their initial size and resource base, and the benefits they derive from the original location of the parent firm.

The technical branch plant takes advantage of support services from the parent firm. Many of these can either by directly provided by the parent or can be secured in the primary center where the parent is located. The bond between the technical branch plant and the parent corporation may therefore reduce clustering or agglomerative tendencies that would otherwise accompany the development of small independent entrepreneurial firms at early stages of the product cycle. The development of technical branch plants thus suggests that the economic structure of new innovation centers is highly influenced by the pre-existence of places such as Silicon Valley and Route 128 and is not independent of them.

Once technical branch plants are established, Glasmeier argues that their products and production processes may facilitate local linkages. Even though they have strong ties to their parents, nevertheless – if the product is customized and made for final users, and exhibits a highly variable input stream with many different components used in irregular quantities – small firms may spring up over time to service their needs. This possibility may, however, be influenced by the development of flexible manufacturing which allows technical branch plants to bring in-house the production of components that might otherwise be provided by local subcontractors and parts suppliers.

The extent of this type of dispersal, made possible by the physical separation of the corporate functions, has recently been studied by Glasmeier (1986a). A principal component analysis of industry occupational employment data for five two-digit high tech industry groups – chemicals, non-electrical equipment, electrical equipment, transportation equipment and scientific instruments (SIC 28, 35, 36, 37, 38) – confirms that technical and administrative occupations are spatially concentrated together, while production and assembly occupations are found in other locations. In addition, Glasmeier also finds that production occupations in the non-electrical and electrical equipment industry groups (SIC 35 and 36) are spatially concentrated in different locations than assembly occupations. She attributes this spatial separation to the distinct labor requirements of firms during the production and assembly phases of high tech manufacturing. Case-study interviews conducted by Saxenian (1981) further substantiate this tendency by showing that firms with headquarters in Santa Clara County had 79% of their research and development facilities, 33% of their advanced manufacturing, and only 3% of their assembly operations located in the Valley. The distance of dispersal also varies with function. Most of the advanced manufacturing plants were located in the non-union West within three hours' plane ride of Silicon Valley, while 80% of the assembly operations were in Third World countries. Saxenian's findings are reproduced in Table 6.1.

Table 6.1 Plant sites of San Jose-based firms, by phase of production, 1980 (percent).

	Control	Advanced R&D	Manufacturing	Assembly
Santa Clara County	100	79	36	3
Pacific Northwest and Southwest	0	0	35	9
rest of United States	0	0	13	0
Europe and Japan	0	21	16	0
Third World	0	0	0	88

Source: Saxenian (1980).

It is this process of concentration and dispersal that our empirical work is aimed at documenting. While our data base, described in the next section, does not distinguish between experimental, advanced manufacturing and mass production phases of the fabrication process, it gives us highly disaggregated data on individual high tech manufacturing industries, permitting an analysis of comparative degrees of dispersion and changes in locational patterns over time.

The locational data base

In order to test these hypotheses about high tech agglomeration and dispersal, and also to make the analyses of detailed location described in Chapters 7 and 9, we assembled an extraordinarily fine-grained data set, which is described in more detail in Appendix 2. It contains plant and employment figures, based on Census of Manufactures data, for each of our 100 industries for all 3,140 counties in the United States. It refers to two Census years, 1972 and 1977; though we hoped to extend our analysis to cover the subsequent 1982 Census, the figures were only just becoming available as we sent this book to press.

Since the Census of Manufactures gives no direct information on employment levels, these have had to be estimated by methods described in Appendix 2. The reader should keep this in mind throughout; though we believe that in the great majority of cases the estimates are close to reality, it is possible that there are errors for some smaller counties. Further, the data base cannot distinguish within an industry between plants with a high proportion of high tech occupations, and those with low proportions; it is always possible that (for instance) a semiconductor plant in Arizona or Colorado is a relatively routine operation, which would not by itself qualify as "high tech." This bias generally works against real centers of innovation and in favor of the outlying areas that are the recipients of the more routine functions. Since many high tech advocates are interested in just this kind of operation, the fact may not matter so much; the reader should simply keep in mind that we are tracking plants in high tech industries, not necessarily ones which are themselves engaged in innovative high tech processes.

Geographical concentration and dispersion

We found significant variation in the degree to which our high tech industries are distributed across space. Some are highly dispersed. Twelve industries, listed in Table 6.2, employed people in over 300 counties. Others were severely restricted in geographical range. Twelve, listed in Table 6.3, provided jobs in less than 50 counties nationwide.

But the number of counties in which plants are sited is only a rough guide to high tech dispersion. As a measure, it does not tell us anything about the relative size of plants. An industry with an apparently broad distributional pattern, signalled by its presence in hundreds of counties, could actually fashion most of its product in a few locations, thereby concentrating jobs quite heavily. For instance, while 117 counties hosted aircraft engine and parts plants in 1977, most of the large plants assembling engines are concentrated in a few counties in the northeastern part of the USA.

As a more precise location indicator of high tech activity, we constructed an "entropy index" for all 100 high tech industries. This index provides a measure of the relative degree of dispersion or concentration in each industry across all 3,140 counties in the USA for the given year.[1] A value of zero implies total spatial dispersion, while a value of 8.16 implies total concentration of the industry in one county.[2] These indices were computed for both plants and employment. Differences between

Table 6.2 Highly dispersed high tech industries.

SIC	Industry name	Incidence by number of counties, 1977
3544	specialty dies, die sets, jigs fixtures, industrial molds	799
3679	electronic components, NEC	521
3569	general industrial machinery equipment, NEC	492
2899	chemicals, chemical preparations, NEC	489
2875	fertilizers, mixing only	469
3662	radio, TV transmitting, signal, detection equipment	456
3531	construction machine equipment	450
2851	paints, varnishes, lacquers, enamels	397
3842	orthopedic, prosthetic, surgical applications	394
2842	special cleaning, polishing preparations	315
2819	industrial inorganic chemicals, NEC	313
2813	industrial gases	309

How high tech clusters

Table 6.3 Highly concentrated high tech industries.

SIC	Industry name	Incidence by number of counties, 1977
3764	guided missiles, space vehicles, propulsion units	19
3795	tanks, tank components	19
3031	reclaimed rubber	20
3761	guided missiles, space vehicles	21
2823	cellulosic man-made fibers	25
2895	carbon black	25
3769	guided missiles, space vehicles, parts, NEC	31
2812	alkalies & chlorine	40
3547	rolling mill machinery equipment	43
3574	calculating accounting machines, except electrical computer equipment	47
3624	carbon, graphite products	48
2822	synthetic rubber	49

plant and job indices enable us to test our hypothesis that jobs are geographically less concentrated than plants. We also computed the indices for both 1972 and 1977. Comparing the two years permits us to detect tendencies toward further dispersion or concentration among high tech sectors during the 1970s.

Inter-industry patterns of dispersal

With the entropy measure, high tech industries show highly variegated degrees of dispersion across US counties (see Table 6.4). In 1977, the industry with the most highly concentrated employment was miscellaneous guided missile and space vehicle parts (SIC 3769). Its nearly 10,000 jobs, sited in just 31 counties, gave it an entropy index value of 6.22. A companion sector, guided missiles and space vehicles (SIC 3761), posted the highest concentration of plants, at 5.38, its 94,000 jobs located in just 21 counties. On the basis of absolute county incidence, the most highly concentrated industry was tanks and tank components (SIC 3795) whose 12,000 plus jobs could be found in only 19, or less than 1%, of all US counties.[3] These defense-oriented industries illustrate that concentration is not simply a function of size, since the assembly of missiles takes place at fewer locations despite the fact that it employs almost ten times as many workers as the missile parts industry.

The industry with the most highly dispersed employment in 1977 as gauged by the entropy index was fertilizers, mixing only (SIC 2875). Its 12,000 jobs were distributed across 469 counties. Its employment

Table 6.4 Entropy indices: a measure of spatial dispersion of high technology industries, 1972 and 1977.

SIC	Industry name	1972 County occur	1972 Entropy indices Plants	1972 Entropy indices Employees	1977 County occur	1977 Entropy indices Plants	1977 Entropy indices Employees
2812	alkalies & chlorine	44	4.29821	5.05486	40	4.46775	5.19241
2813	industrial gases	274	2.69220	3.19407	309	2.56153	3.09774
2816	inorganic pigments	76	3.89467	4.68663	72	3.94716	4.54590
2819	industrial inorganic chemicals, NEC	236	2.81602	3.73013	313	2.63022	3.73827
2821	plastic materials, synthetic resins	181	3.20180	3.78313	209	3.10118	3.60944
2822	synthetic rubber	43	4.45826	5.64618	49	4.26684	5.55025
2823	cellulosic man-made fibers	18	5.16352	5.69092	25	4.83501	5.90025
2824	synthetic organic fibers, except cellulose	55	4.07937	4.60663	62	3.94825	4.43842
2831	biological products	118	3.47977	4.65618	188	3.06695	4.23139
2833	medical, chemical, botanical products	82	3.99089	5.14966	111	3.64047	4.84810
2834	pharmaceutical preparations	247	3.15630	3.96218	265	3.08370	3.85240
2841	soap, other detergents	213	3.31717	4.13965	226	3.27265	3.96806
2842	special cleaning, polishing preparations	336	2.98038	3.77162	315	3.04413	3.71424
2843	surface active finishing agents	85	4.04028	4.41353	82	4.03060	4.38446
2844	perfumes, cosmetics, toilet preparations	181	3.80619	4.25820	195	3.68899	4.27868
2851	paints, varnishes, lacquers, enamels	366	3.08434	3.56771	397	2.93345	3.42728
2861	gum, wood chemicals	109	3.46736	4.79593	92	3.66809	4.48424
2865	cyclic crudes, intermediates, dyes	105	3.72751	4.21191	112	3.62403	4.28031
2869	industrial organic chemicals, NEC	229	3.01904	3.91762	255	2.96788	3.81098
2873	nitrogenous fertilizers	64	3.94868	4.29060	123	3.31962	3.98633
2874	phosphatic fertilizers	110	3.50210	4.34658	73	3.94467	4.91344
2875	fertilizers, mixing only	439	2.12409	2.69135	469	2.07958	2.65383
2879	pesticides, agricultural chemicals, NEC	243	2.82687	3.89104	260	2.76536	4.03603

SIC	Description						
2891	adhesives, sealants	171	3.45453	3.91834	211	3.27477	3.66840
2892	explosives	72	3.88067	4.98076	78	3.79270	4.70276
2893	printing ink	119	3.82740	4.23783	145	3.62907	4.13801
2895	carbon black	29	4.77099	5.16842	25	4.92197	5.08589
2899	chemicals, chemical preparations, NEC	438	2.78919	3.31870	489	2.60660	3.07105
2911	petroleum refining	187	3.11418	4.03319	198	3.07332	3.98936
3031	reclaimed rubber	17	5.33541	6.13662	20	5.07538	5.98595
3482	small arms ammunition	52	4.16723	5.66444	52	4.19419	5.81394
3483	ammunition, except small arms, NEC	72	4.02921	4.62404	68	3.93554	4.76640
3484	small arms	56	4.28523	5.48365	72	4.05189	5.44448
3489	ordnance, accessories, NEC	64	3.98185	5.37227	66	4.05380	5.23738
3511	steam, gas, hydraulic turbines	59	4.07156	5.18595	66	3.96583	5.21701
3519	internal combustion engines, NEC	113	3.65636	4.54103	137	3.50746	4.52685
3531	construction machine equipment	395	2.49277	3.58671	450	2.36011	3.51794
3532	mining machinery equipment	158	3.17442	4.13969	189	3.06122	3.85607
3533	oilfield machinery equipment	100	4.48193	5.47295	145	4.21303	5.28409
3534	elevators, moving stairways	92	3.77435	4.57928	97	3.68673	4.40429
3535	conveyors, conveying equipment	245	2.99351	3.57346	284	2.83000	3.25528
3536	hoists, industrial cranes, monorail systems	119	3.52187	4.55385	152	3.28858	3.99822
3537	industrial trucks, tractors, trailers, stackers	230	2.94219	4.20469	251	2.94692	4.03463
3541	machine tools, metal cutting types	242	3.47124	4.22301	277	3.30345	4.13432
3542	machine tools, metal forming types	169	3.55513	4.27489	195	3.31188	4.18921
3544	specialty dies, die sets, jigs fixtures, industrial molds	700	2.93051	3.21653	799	2.78421	3.11511
3545	machine tool accessories, measuring devices	292	3.46002	3.85542	350	3.33277	3.75678
3546	power driven hand tools	59	4.29943	4.89581	90	3.76883	4.57621
3547	rolling mill machinery equipment	34	4.63894	5.57224	43	4.52980	5.43706
3549	metalworking machinery, NEC	182	3.37909	3.78980	250	3.02589	3.46931
3561	pumps, pumping equipment	253	2.97831	3.54232	286	2.88280	3.49868
3562	ball, roller bearings	86	3.81680	4.48919	102	3.63955	4.30788

Table 6.4 Entropy indices: a measure of spatial dispersion of high technology industries, 1972 and 1977 – continued

| | | | 1972 | | | 1977 | |
| | | | Entropy indices | | | Entropy indices | |
SIC	Industry name	County occur	Plants	Employees	County occur	Plants	Employees
3563	air, gas compressors	61	4.12440	4.93378	105	3.71224	4.53192
3564	blowers, exhaust, ventilation fans	185	3.18544	3.81928	231	3.00482	3.53355
3565	industrial patterns	259	3.11612	3.51042	286	3.01183	3.36969
3566	speed changers, industrial high drives, gears	142	3.59308	4.48322	139	3.53357	4.33501
3567	industrial process furnaces, ovens	120	3.65385	4.26595	137	3.57044	4.13794
3568	mechanical power transmission equipment, NEC	105	3.59098	4.30585	140	3.34673	4.00528
3569	general industrial machinery equipment, NEC	311	2.96574	3.36581	492	2.66028	3.04639
3573	electronic computing equipment	157	3.82266	4.25033	203	3.69157	4.10469
3574	calculating accounting machines, except electrical computer equipment	57	4.15283	5.35940	47	4.38367	5.31164
3576	scales, balances, except laboratory	68	4.08261	5.07204	78	3.88224	4.71723
3579	office machinery, NEC	111	3.72568	4.78868	112	3.71896	4.86564
3612	power, distribution special transformers	148	3.33270	4.24433	175	3.17599	4.01271
3613	switch gear, switchboard apparatus	217	3.18153	3.96372	262	3.01880	3.70524
3621	motors, generators	231	3.06761	3.43955	248	2.95124	3.30834
3622	industrial controls	215	3.18969	4.29268	254	3.06443	3.85358
3623	welding apparatus, electric	86	3.99431	4.57242	97	3.84317	4.67045
3624	carbon, graphite products	50	4.32432	5.44934	48	4.38193	5.28373
3629	electrical industrial apparatus, NEC	123	3.79006	4.44025	126	3.62306	4.32572
3651	radio, TV receiving sets, except communication types	157	3.67572	4.61421	220	3.34809	4.38946
3652	phono records, pre-recorded magnetic tape	153	4.35035	4.72780	182	4.30354	4.69896
3661	telephone, telegraph apparatus	121	3.56835	4.37739	144	3.56011	4.37823
3662	radio, TV transmitting, signal, detection equipment	392	3.14078	3.68679	456	3.00989	3.60784

SIC	Description						
3671	cathode ray tubes, NEC	24	4.89046	5.93510	93	3.78179	4.87745
3674	semiconductors, related devices	120	3.94761	4.72746	182	3.71375	4.82531
3675	electronic capacitors	78	3.97069	4.21569	69	4.22650	4.40286
3676	resistors for electronic applications	*	*	*	41	4.05231	4.67983
3677	resistors, electric apparatus	128	3.72759	3.81388	151	3.57894	3.92570
3678	connectors, electronic applications	51	4.54095	4.96639	72	4.19741	4.65758
3679	electronic components, NEC	387	3.14655	3.38748	521	3.03019	3.37465
3721	aircraft	101	3.91201	5.17199	101	3.98548	5.11358
3724	aircraft engines, parts	99	4.04402	4.86717	117	3.86922	4.75432
3728	aircraft parts, auxiliary equipment, NEC	175	4.15420	4.51623	210	4.01012	4.56343
3743	railroad equipment	104	3.80563	4.71564	128	3.51952	4.68313
3761	guided missiles, space vehicles	34	4.88427	5.61447	21	5.37594	5.72779
3764	guided missiles, space vehicles, propulsion units	23	5.03926	5.51446	19	5.22234	5.58615
3769	guided missiles, space vehicles, parts, NEC	31	4.94557	5.54201	31	4.85679	6.21897
3795	tanks, tank components	17	5.32548	6.40121	19	5.20825	6.10139
3811	engineering, laboratory, scientific, research instruments	227	3.25312	3.84317	259	3.11617	3.80239
3822	industrial controls for regulators, resistors, communications and environmental applications	91	3.78759	4.91342	129	3.49726	4.71313
3823	industrial instruments for measurement and display	88	3.97757	4.86775	161	3.54257	4.41280
3824	fluid meters, counting devices	49	4.28691	4.85434	81	3.79483	4.79823
3825	instruments, measuring, testing, electrical, electrical signals	211	3.37575	4.09865	230	3.29313	4.02826
3829	measuring, controlling devices, NEC	208	3.34942	4.42026	233	3.20060	4.25678
3832	optical instruments, lenses	171	3.58471	4.37220	198	3.38959	4.16984
3841	surgical, medical instruments apparatus	202	3.32506	3.77002	240	3.21986	3.73401
3842	orthopedic, prosthetic, surgical applications	305	2.93260	3.57749	394	2.72335	3.49959
3843	dental equipment, supplies	162	3.60521	4.24015	208	3.31918	4.11538
3861	photographic equipment, supplies	187	3.69423	5.13158	225	3.50611	4.79331

*For technical reasons the entropy index for this industry cannot be calculated for 1972.

dispersion index value was 2.65 and plant value was 2.08; both were the lowest in the set. On a county-by-county basis, specialty dies and industrial molds (SIC 3544) could be found in 799 counties (or 25%), and miscellaneous electronic components (SIC 3679) in 521 (or 17%). However, in both these sectors the distributions of plants and jobs were highly skewed among these few counties, so that they exhibited less real dispersion with the entropy index than did fertilizers.

Several generalizations can be made about factors causing this wide range of geographical distribution. First, as illustrated above with missiles and space equipment, there appears to be a rather weak relationship between the level of employment in an industry and its tendency to disperse. Some very large employers are highly concentrated, while other sectors with only modest national levels of employments are quite dispersed. Scale economies within sectors appear to play a much greater role in the tendency for some industries to concentrate, irrespective of total workforce size.

To test the hypothesis that industry dispersion is more closely related to plant scale than to absolute size, we ran a set of regressions.[4] Using all 100 industries as observations, we did find modest negative relationships between workforce, plant numbers and dispersion values. However, we could explain only 5% of the variation in job indices and 18% of that for plant indices. The exercise also suggests that a large number of plants, and hence low average plant size, are more closely associated with geographical spreading than a large national workforce.

Secondly, industries appear to concentrate or disperse by type of product, major customer and resource dependency. For instance, two types of industries are prominent among the concentrators. Defense and space-oriented high tech manufacturers are extraordinarily concentrated. Of the 18 sectors with job entropy values above 5.0 in 1977, 8 produce weapons, tanks, spacecraft, airplanes and related parts. It is difficult not to draw the conclusion that military and space spending patterns and procurement practices have been a major contributor to the spatial concentration of an important segment of high tech sectors in a small number of counties. Whether this represents economies of scale alone or politically influenced procurement practices in addition is not clear. It may be, as critics have suggested, that Congressional pork barrelling, presidential preferences (Johnson's Gulf-oriented space program) and Pentagon "geographical constituency building" may have contributed to this pattern.[5]

A quite different set of highly resource-oriented industries are also heavily concentrated geographically. These sectors account for another 7 of the top 18 highly concentrated industries and run the gamut from oilfield machinery to raw-materials-related commodities such as carbon products, alkalies and chlorine, synthetic rubber and man-made fibers. In some cases, such as oilfield machinery, it is *demand* from a resource sector that has drawn the industry to a small set of locations. In others, it is the *supply* of a raw material or a primary processing sector input which accounts for its concentration. In these latter cases, the traditional location theory arguments about weight-shedding activities appear to dictate plant sitings.

In both these two types of sectors, the "age" of the industry appears to have less to do with its locational concentration than its product type and client relationships. In addition, plants in these industries are often quite large scale in nature, either because their product is immense (aircraft, missiles) or because continuous process technology facilitates very large economies of scale. Therefore, a few big plants serve most of the market.

At the other end of the spectrum, a set of relatively mature producer goods sectors dominate the rankings. Many of these highly dispersed sectors produce heavy or bulky material inputs for a relatively dispersed set of industrial or agricultural consumers (fertilizers, paints, chemicals). Others produce custom-made equipment for relatively dispersed industrial users (dyes, motors, electronic components, conveyors). In both instances, market orientation seems to be drawing the industry.

In the middle range lie many of those high tech industries which are innovative and fastest growing. Computers, semiconductors, biological products, measuring devices, industrial controls, optical instruments, and machine tools are all only moderately dispersed compared to the average for all high tech industries. This may be a function of youth and the need to cluster in certain places in order to watch competitors, draw upon secondary business services and a skilled labor pool, and be close to the centers of action. Yet it is still true that semiconductors could be found in a total of 182 counties in 1977, and computers in 203, suggesting that decentralization of at least some innovative high tech production activities was fairly advanced.

Changes in patterns of concentration, 1972–7

Over the period studied, there has been strong tendency for both plants and employment to disperse across US counties. In only 14 of our 100 industries (Table 6.4) did the plant entropy index increase, indicating cases of greater concentration. The job entropy index increased in 19, or about one in five, sectors. If anything, the degree of differentiation among industries increased. Almost universally, dispersed sectors became even more ubiquitous while the highly concentrated ones were as apt to intensify their plant concentrations as they were to disperse.

More specifically, the vast majority of sectors which did increase their plant concentrations fall into the two groups mentioned above: military-related sectors and bulk materials processors. Here, the forces originally working toward concentration seem to be continuing that trend. This is particularly striking in the military equipment and supplies case, where small arms ammunition, guided missiles, space vehicles and parts, aircraft, and aircraft parts were found among the minority of sectors that intensified either plant or job concentrations.

Generally, the more commercially oriented innovative high tech sectors tended to disperse along with other sectors. However, variations within these groups appear to be greater than for other groups. In a few cases, employment continued to concentrate at existing sites, supporting a product cycle interpretation which hypothesizes concentration in initial stages. Semiconductors is an outstanding example. While plants dispersed in the 1970s, and the industry increased its incidence from 120 to 182 counties, jobs actually became more concentrated geographically. Several other electronic sectors also became more concentrated (resistors, telephone equipment). Most other rapid growth sectors showed a modest tendency toward decentralization from original growth centers.

The only other sectors showing a tendency to increase locational concentration rather than to disperse were a set of relatively highly automated process sectors, whose initial locations were already relatively dispersed. Examples are industrial inorganic chemicals, pesticides and agricultural chemicals, cosmetics, phosphatic fertilizers, and finishing agents. The tendency to reconcentrate here may often be a result of rationalization – selective plant closings and relatively larger-scale plants in new or existing locations.

The geographical redistribution of high tech industry

Product–profit cycle theory contends that redistribution of employment is the outcome of the tendency to disperse. We computed another set of coefficients to capture the relative degree of redistribution among our high tech constituents in the period 1972–7. The coefficient of redistribution, R, is computed in the following manner:

$$\sum_{r=1}^{3140} \left| \frac{E_{ir}^{1972}}{E_i^{1972}} - \frac{E_{ir}^{1977}}{E_i^{1977}} \right|$$

where E_i is the national employment in industry i, and E_{ir} is the county employment in industry i. It is a measure of the deviation between two distributions of the same phenomenon taken at these two points in time. The value of the coefficient ranges from 0 to 1; zero implies no redistribution whereas 1 implies complete redistribution.

As in the dispersion issue, high tech industries showed substantial idiosyncrasy in redistributional tendencies in the 1970s (see Table 6.5). Those which showed the greatest shifts are a rather disparate lot. Shown in Table 6.6 in descending order, they are headed by cathode ray tubes (SIC 3671), guided missile and space vehicle parts (3769), calculating machines other than computers (3574), and biological products (2831). In all of these industries, employment levels changed by more than 20% in the period, affording each the opportunity of geographical restructuring. This "potentially mobile employment", in the phrase coined by Massey and Meegan (1982), may or may not result in geographical redistribution. In the top three on our list, employment actually declined by 20%, 66% and 24% respectively, so that the coefficients reflect the fact that these cuts are very unevenly distributed across space. In the biological products industry, an increase of 55% was also heavily skewed. In the cases of job loss, it is not possible to tell from the coefficient whether reconcentration is taking place at old or new sites, whereas in the job gain industries, a large coefficient of redistribution clearly indicates shifts toward new locations.

Job redistribution is not simply a matter of high absolute or relative levels of potentially mobile employment. A number of industries which posted large gains and losses in the mid-1970s exhibit relatively even geographical distribution of these changes, reflected in relatively low coefficients. For instance, both the tank and tank components industry (SIC 3795) and the oilfield machinery industry (3533) increased employment dramatically from 1972 to 1977, by 110% and 63% respectively. Yet both evidently expanded their operations in situ, resulting in little spatial repatterning of the industry. Among heavy job-losing industries, small arms ammunition (3482) and explosives (2892) appear to have taken their cuts relatively evenly. And in some cases of

Table 6.5 County and city locations entropy index and coefficient of redistribution, 1977.

SIC	Industry name	County occurrence	SMSA occurrence	Employment entropy index	Employment redistribution index
3769	guided missiles, space vehicles, parts, NEC	31	26	6.21897	0.73430
3795	tanks, tank components	19	12	6.10139	0.29866
3031	reclaimed rubber	20	14	5.98595	0.29974
2823	cellulosic man-made fibers	25	14	5.90025	0.36022
3482	small arms ammunition	52	32	5.81394	0.28078
3761	guided missiles, space vehicles	21	18	5.72779	0.28030
3764	guided missiles, space vehicles, propulsion units	19	18	5.58615	0.39480
2822	synthetic rubber	49	37	5.55025	0.21996
3484	small arms	72	49	5.44448	0.18279
3547	rolling mill machinery equipment	43	25	5.43706	0.38620
3574	calculating accounting machines, except electrical computer equipment	47	36	5.31164	0.67710
3533	oilfield machinery equipment	145	62	5.28409	0.21806
3624	carbon, graphite products	48	28	5.28373	0.16077
3489	ordnance, accessories, NEC	66	42	5.25738	0.27117
3511	steam, gas, hydraulic turbines	66	46	5.21701	0.20243
2812	alkalies & chlorine	40	24	5.19241	0.22430
3721	aircraft	101	55	5.11358	0.17883
2895	carbon black	25	10	5.08589	0.14697
2874	phosphatic fertilizers	73	35	4.91344	0.37143
3671	cathode ray tubes, NEC	93	67	4.87745	0.88509
3579	office machinery, NEC	112	60	4.86564	0.33906
2833	medical, chemical, botanical products	111	70	4.84810	0.39715
3674	semiconductors, related devices	182	96	4.82531	0.30290
3824	fluid meters, counting devices	81	53	4.79823	0.49038
3861	photographic equipment, supplies	225	121	4.79331	0.28936
3483	ammunition, except small arms, NEC	68	42	4.76640	0.39612
3724	aircraft engines, parts	117	69	4.75432	0.23419
3576	scales, balances, except laboratory	78	47	4.71723	0.32180
3822	industrial controls for regulators, resistors, communications and environmental applications	129	76	4.71313	0.37228
2892	explosives	78	34	4.70276	0.31475
3652	phono records, pre-recorded magnetic tape	182	100	4.69896	0.23009
3743	railroad equipment	128	63	4.68313	0.21110
3623	welding apparatus, electric	97	58	4.67045	0.20969
3678	connectors, electronic applications	72	37	4.65758	0.32548
3546	power driven hand tools	90	57	4.57621	0.33710
3728	aircraft parts, auxiliary equipment, NEC	210	95	4.56343	0.27662
2816	inorganic pigments	72	38	4.54590	0.26188
3563	air, gas compressors	105	60	4.53192	0.28969
3519	internal combustion engines, NEC	137	69	4.52685	0.26991
2861	gum, wood chemicals	92	24	4.48424	0.42325
2824	synthetic organic fibers, except cellulose	62	31	4.43842	0.23994

Table 6.5 County and city locations entropy index and coefficient of redistribution, 1977 – *continued*.

SIC	Industry name	County occurrence	SMSA occurrence	Employment entropy index	Employment redistribution index
3823	industrial instruments for measurement and display	161	92	4.41280	0.31160
3534	elevators, moving stairways	97	63	4.40429	0.37950
3675	electronic capacitors	69	35	4.40286	0.29606
3651	radio, TV receiving sets, except communication types	220	108	4.38946	0.42482
2843	surface active finishing agents	82	46	4.38446	0.29102
3661	telephone, telegraph apparatus	144	72	4.37823	0.23161
3566	speed changers, industrial high drives, gears	139	76	4.33501	0.26515
3629	electrical industrial apparatus, NEC	126	69	4.32572	0.31236
3562	ball, roller bearings	102	48	4.30788	0.16948
2865	cyclic crudes, intermediates, dyes	112	63	4.28031	0.31034
2844	perfumes, cosmetics, toilet preparations	195	101	4.27868	0.20452
3829	measuring, controlling devices, NEC	233	119	4.25678	0.28366
2831	biological products	188	108	4.23139	0.60489
3542	machine tools, metal forming types	195	99	4.18921	0.20446
3832	optical instruments, lenses	198	106	4.16984	0.38631
2893	printing ink	145	76	4.13801	0.16976
3567	industrial process furnaces, ovens	137	73	4.13794	0.24784
3541	machine tools, metal cutting types	277	118	4.13432	0.18651
3843	dental equipment, supplies	208	114	4.11538	0.31484
3573	electronic computing equipment	203	109	4.10469	0.32466
3676	resistors for electronic applications	41	3	4.05231	*
2879	pesticides, agricultural chemicals, NEC	260	127	4.03603	0.44213
3537	industrial trucks, tractors, trailers, stackers	251	114	4.03463	0.33547
3825	instruments, measuring, testing, electrical, electrical signals	230	119	4.02826	0.25995
3612	power, distribution special transformers	175	83	4.01271	0.30857
3568	mechanical power transmission equipment, NEC	140	75	4.00528	0.30009
3536	hoists, industrial cranes, monorail systems	152	83	3.99822	0.40765
2911	petroleum refining	198	88	3.98936	0.13363
2873	nitrogenous fertilizers	123	61	3.98633	0.29609
2841	soap, other detergents	226	119	3.96806	0.14856
3677	resistors, electric apparatus	151	68	3.92570	0.44259
3532	mining machinery equipment	189	82	3.85607	0.34359
3622	industrial controls	254	119	3.85358	0.31232
2834	pharmaceutical preparations	265	131	3.85240	0.20814
2869	industrial organic chemicals, NEC	255	118	3.81098	0.25518
3811	engineering, laboratory, scientific, research instruments	259	121	3.80239	0.30359
3545	machine tool accessories, measuring devices	350	140	3.75678	0.19816
2819	industrial inorganic chemicals, NEC	313	126	3.73827	0.23404

Table 6.5 County and city locations entropy index and coefficient of redistribution, 1977 – *continued.*

SIC	Industry name	County occurrence	SMSA occurrence	Employment entropy index	Employment redistribution index
3841	surgical, medical instruments apparatus	240	123	3.73401	0.38533
2842	special cleaning, polishing preparations	315	150	3.71424	0.26082
3613	switch gear, switchboard apparatus	262	118	3.70524	0.22517
2891	adhesives, sealants	211	104	3.66840	0.31549
2821	plastic materials, synthetic resins	209	94	3.60944	0.25464
3662	radio, TV transmitting, signal, detection equipment	456	183	3.60784	0.22965
3564	blowers, exhaust, ventilation fans	231	122	3.53355	0.27958
3531	construction machine equipment	450	167	3.51794	0.18318
3842	orthopedic, prosthetic, surgical applications	394	188	3.49959	0.30847
3561	pumps, pumping equipment	286	133	3.49868	0.18766
3549	metalworking machinery, NEC	250	114	3.46931	0.32132
2851	paints, varnishes, lacquers, enamels	397	179	3.42728	0.14470
3679	electronic components, NEC	521	187	3.37465	0.30866
3565	industrial patterns	286	132	3.36969	0.21355
3621	motors, generators	248	106	3.30834	0.22988
3535	conveyors, conveying equipment	284	111	3.25528	0.31757
3544	specialty dyes, die sets, jigs fixtures, industrial molds	799	209	3.11511	0.13738
2813	industrial gases	309	157	3.09774	0.32520
2899	chemicals, chemical preparations, NEC	489	184	3.07105	0.28466
3569	general industrial machinery equipment, NEC	492	176	3.04639	0.34638
2875	fertilizers, mixing only	469	136	2.65383	0.35637

*For technical reasons the employment redistribution index cannot be calculated.

modest job change, such as the hoists, industrial cranes and monorail systems industry (3536), significant redistribution did take place in the period, indicating that geographical shifts are not simply a counterpart to disruptive growth patterns.

Nevertheless, there is a strong correspondence between the availability of potentially mobile employment (and unemployment) and observed changes in the geographical pattern of distribution. This indicates to us that high tech industries in this period did take advantage of growth or shrinkage imperatives to reorient themselves spatially. Dispersion does appear to be strongly associated with job growth. It is not possible to decipher whether the observed redistribution in job loss industries represents movement to new locations or reconcentration in older centers. But we *can* reject the hypothesis that job gains and cuts are more or less evenly distributed proportionately across the nation's counties.

How high tech clusters 93

Table 6.6 Ranking redistributive high tech sectors

SIC	Industry name	Coefficient of redistribution
3671	cathode ray tubes, NEC	0.88509
3769	guided missiles, space vehicles, parts, NEC	0.73430
3574	calculating accounting machines, except electrical computer equipment	0.67710
2831	biological products	0.60489
3824	fluid meters, counting devices	0.23419
3677	resistors, electric apparatus	0.44259
2879	pesticides, agricultural chemicals, NEC	0.44213
3651	radio, TV receiving sets, except communication types	0.42482
2861	gum, wood chemicals	0.42325
3536	hoists, industrial cranes, monorail systems	0.40765
2833	medical, chemical, botanical products	0.39715
3483	ammunition, except small arms, NEC	0.39612
3764	guided missiles, space vehicles, propulsion units	0.39480
3832	optical instruments, lenses	0.38631
3547	rolling mill machinery equipment	0.38620
3841	surgical, medical instruments apparatus	0.38533

Some unresolved questions

A number of intriguing issues remain. One, which we were not able to pursue, is the degree to which industries are geographically associated. Industrial complex analysis would lead us to suppose that certain industries are really the key to high tech tendencies toward concentration and dispersal. Measures of geographical association, and tests of probability that industry will be found in conjunction with industry geographically, would offer some insight into these relationships.

A second issue concerns the degree to which industry shifts at the county level, detected without coefficients, reflect central city to suburban, or urban to rural, moves within a region as opposed to interregional movements of jobs and plants. As it stands, our findings on dispersion do not distinguish between these types of movement. We do know that industries vary in the degree to which they cluster within certain SMSAs; some industries, especially market-oriented ones, will tend to sport few plants in a SMSA whereas an industry that is concentrated in a few SMSAs may be found in several of their constituent counties. Our measures of dispersion might rank two such disparate industries more or less equally, because they could be found in a similar number of counties.

Some evidence on this problem can be gleaned from Table 6.5, which shows the number of counties and the number of SMSAs in which each industry is sited. An extreme case is represented by guided missile and space vehicle propulsion units (SIC 3764) plants, which are to be found in 19 counties representing 18 SMSAs. The data in this form do not tell us whether the difference is accounted for by two engine plants located in two different counties within one SMSA, or a plant sited in a non-SMSA county. To illustrate with the other extreme, three industries – gum and wood chemicals (2861), specialty dyes and industrial molds (3544) and fertilizers (2875) – all are found in more than three times as many counties as SMSAs. In the case of the tool and die industry, this is probably because plants can be found in tandem in most of the constituent counties in certain SMSAs like Detroit and Chicago. But in the other two industries, which are heavily resource and rural oriented, the difference must be accounted for by heavy concentrations of plants in non-metropolitan counties.

One final and related issue: we were not able to test the sensitivity of our results to the heterogeneity of county size. In some parts of the country, counties are relatively small in relation to others, so that some bias may enter into the numbers from purely jurisdictional sources. In the West, counties tend to be immense, with only one or two comprising an SMSA. In contrast, eastern and midwestern cities tend to have relatively smaller counties with a larger average number in each SMSA. Of course, population densities are also greater on average in the latter, but counties remain a relatively imperfect geographical category for this reason. Industries with a preponderance of plants in large, western counties may appear to be more geographically concentrated than their counterparts in eastern metropolitan areas, even though the actual distances separating facilities may be the same. It would be interesting in general to pursue this issue of differing county size and its effect on county-level distributional data.

Summary

In the first portion of this chapter, we presented a theoretical elaboration of our earlier work on geographical concentration and dispersion in high tech industries, along with some anecdotal examples. We argued that nascent high tech industries will tend to agglomerate in a few locations, and will begin to decentralize activities only when products have become relatively standardized. Our findings indicated that tendencies toward decentralization and redistribution of plants and employment were quite common in the 1970s, especially for those industries with severe growth or shrinkage pains.

Not much evidence of tendencies toward exclusive agglomeration was detected, even in relatively youthful and generally acknowledged leading edge industries like computers and semiconductors. Semiconductor plants, for instance, could be found in 182 counties in 1977 compared to 120 in 1972. Yet the employment entropy index for semiconductors actually increased over the period, and we know from other sources that California increased its share of semiconductor plants in the USA consistently over the period 1963–77 (Markusen 1985a). This suggests that in this industry, until now, dispersion may be more a metropolitan and intraregional phenomenon rather than interregional. Similarly, California increased its share of computer plants in the period 1967–77, even though nationally computer plants were springing up in increasing numbers of counties.

Two factors account for the relative absence of observed increases in agglomeration in our set of high tech industries over time. The first is that many of the industries studied have already undergone their initial periods of concentration. Indeed, by the time an industry is established enough to receive an SIC code, its progress through the profit cycle is already relatively advanced. Computers, for instance, only received an SIC designation in 1967 and semiconductors in 1963; both experienced their initial periods of intense, innovative activity in the decade or so before these dates. The modest levels of dispersion detected with our measures reflect the beginnings of the process of market penetration and cost cutting.

Secondly, certain secular changes in business organization have speeded up the process of dispersion. The emergence of the modern corporation, with its ability to segregate functions geographically, and the construction of an elaborate system of information on site possibilities in specialized firms such as the Fantus Corporation, encourage even youthful industries to take advantage of geographical differences. As high tech firms are swallowed up by conglomerates with tremendous financial, marketing and planning resources, their managements can take advantage of internal efficiencies which previously took many decades for new sectors to generate.

We must conclude, therefore, that there is much potentially mobile employment in high tech sectors, particularly those that are growing rapidly. Industries posting big job gains in the 1970s did demonstrate a will to move, to decentralize geographically and to redistribute employment and plants. This is good news for places that have to date missed out on high tech economic development. Not all of the jobs in industries classified as high tech will remain in centers with a head start. But, on the other hand, not just any locality can expect to attract high tech jobs. Indeed, as the next chapter shows, the distribution is highly uneven over regions, states and metropolitan areas. Just who does get high tech jobs, and why, are the subjects of the rest of the book.

Notes

1. The entropy index is calculated by the following formula:

$$I(i) = \sum_{r=1}^{R} y_{ri} \log \frac{y_{ri}}{1/R}$$

where $I(i)$ is the overall US entropy index of geographical inequality for industry i, y is the share of the rth county of all US employment in industry i, and R is the total number of counties in the US ($R = 3,140$). The entropy index was first developed by Theil (1967) and has been used by locational researchers, primarily in Britain (Keeble 1976, Martin 1972, Chisholm & Oeppen 1973).

2. The absolute values of the indices are less interesting than their comparative size across industries and changes in them over time. We would not expect all industries to be equally distributed across US counties and, not surprisingly, the most ubiquitous of the set was to be found in only 799 out of the 3,140 counties in 1977, while second place went to an industry found in 521 counties.

3. The tank industry is an example of how an industry which is found in only a few counties can at the same time be more dispersed among those counties, because only one or two plants are found in each, as compared with another industry such as missile parts.

4. Entropy indices for both plants and jobs for both years were treated as the dependent variables, absolute numbers of jobs and plants as independent variables. The estimates, their significance levels, and R^2 were as follows:

	α	β		
$I_i^{77E} =$	4.51 +	− (0.000003) E_i^{77}	0.05	5%
$I_i^{72E} =$	4.61 +	− (0.000003) E_i^{72}	0.05	5%
$I_i^{77P} =$	3.75 +	− (0.00033) P_i^{77}	0.18	1%
$I_i^{72P} =$	3.88 +	− (0.00036) P_i^{72}	0.18	1%

where E is employment, P is plant, and $77E$ employment in 1977, etc.

5. See Markusen (1986) for an elaboration of some of these points.

7

Where high tech locates

Regional and urban distribution patterns

IT IS NOW the time to focus on the places where high tech locates. If it is true – as Chapter 6 has suggested – that innovative high tech industries agglomerate in early stages of their evolution, it should be possible to identify these centers. This is the first task of this chapter. We do so at both state and metropolitan scales. In addition, we investigate the patterns of dispersion from these centers over the mid-1970s period, to establish whether the dominant tendency is intraregional dispersion or shifts between major regions. Our findings suggest that there are five major regional agglomerations of high tech industry in the USA, and five smaller ones of note. Within these, there are some cases of multiple nodes of agglomeration. In most of the case areas, agglomeration continued in the 1970s. But for the most part, the major centers grew more slowly than both their own peripheries and the smaller, detached centers. At the same time, Sunbelt/Frostbelt differences overlaid the process, worsening the performance of the more northeasterly agglomerations compared to those in the south and west.

Regional high tech agglomerations
the core high tech states

Agglomeration can exist at different geographical scales. It can occur on an extensive, regional scale, reflecting the transportation, financial, and political externalities within a wide territorial area. Or it can operate more narrowly, referring to labor market and business services proximity found in metropolitan areas. Here, we look at both types. The former we call "regional agglomerations" and the latter "metropolitan agglomerations." Since the latter are embedded in the former, we first identify the major regional centers and subsequently look at the metropolitan agglomerations within them.

We have chosen to use states as the basis for sketching out the larger agglomerations, even though they only crudely capture the nodal and/or homogenous qualities sought in delineating regions. The four umbrella regions – Northeast, North Central, South, and West – are too large for our purposes, and the nine Census divisions into which they are broken down create boundaries that in too many instances artificially cordon off one core high tech state from its real regional partners.[1] While we do present the aggregate figures for these larger groupings, the state-level analysis permits us to compare in a finer-grained manner each lead state with its periphery, and has the added advantage that most readers are quite familiar with the area encompassed within each state. That familiarity will also help readers mentally adjust for anomalies in state size and internal homogeneity.

Agglomeration represents, in a rather complex fashion, both sheer size and specialization. In order to identify states which are high tech centers in both senses, we looked at two indicators. First, we looked at sheer size: numbers of high tech jobs accounted for by states in 1977. Secondly, we computed location quotients for each state, in which the state's share of high tech employment nationally is compared to its share of national manufacturing employment.[2] This indicator prevents us from classifying a state's economy as a high tech agglomeration simply because it has a high absolute level of high tech jobs, which can be simply because it is a big state. New York is a good example. With 337,000 high tech jobs, it is actually the third-ranked state by numbers of jobs. Yet because it has no relative concentration of high tech, it fails to gain inclusion in the major high tech concentration based on New Jersey and Maryland.

Leading states according to both criteria are listed in Table 7.1. The top

Table 7.1 Leading high tech states, 1977.

	By absolute high tech employment		
Rank	State	Jobs (000)	Location quotient
1	California	641.3	1.49
2	Illinois	360.3	1.15
3	New York	336.8	0.89
4	Pennsylvania	314.3	0.94
5	Ohio	295.1	0.88
6	Texas	285.7	1.33
7	New Jersey	232.3	1.23
8	Massachusetts	204.6	1.43
9	Michigan	169.4	0.86
10	Connecticut	160.0	1.65
11	Indiana	153.5	0.89
12	Wisconsin	128.8	1.00
13	Florida	119.0	1.39

	By location quotient		
Rank	State	Location quotient	Jobs (000)
1	Arizona	1.80	45.9
2	Connecticut	1.65	160.0
3	Kansas	1.63	62.8
4	Maryland	1.60	48.5
5	Colorado	1.57	52.9
6	California	1.49	641.3
7	Massachusetts	1.43	204.6
8	Florida	1.39	119.0
9	Oklahoma	1.38	53.9
10	Texas	1.33	285.7
11	Utah	1.30	22.9
12	New Jersey	1.23	232.3
13	Louisiana	1.22	56.4

Source: Glasmeier (1986b).

five states according to the first criterion (California, Illinois, New York, Pennsylvania, and Ohio) are a completely different set from those which have relatively high levels of manufacturing specialization in high tech (Arizona, Connecticut, Kansas, Maryland and Colorado; see Fig. 7.1). Indeed, such states as New York and Pennsylvania have location quotients less than 1, indicating that high tech industries are actually under-represented in their industrial complexes. States like these cannot be considered high tech agglomerations; they are general manufacturing agglomerations which have hosted high tech growth, but not in proportion to the sector's importance nationally.

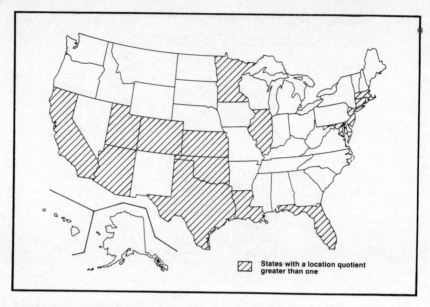

Figure 7.1 High tech employment: location quotients, 1977. Minnesota and Illinois are included here but do not appear on Table 7.6.

To identify major agglomerations from these two indicators, we looked for contiguous groupings of states with similar high rankings on the specialization scale and relatively high levels of high tech employment. Five major agglomerations stand out – neighboring or single states accounting for more than 300,000 high tech jobs and with location quotients greater than 1 (Table 7.2 and Fig. 7.2).

A second set of regional centers, what we will call minor high tech cores, consists of single states whose neighbors are generally not similarly specialized, and whose manufacturing jobs are skewed toward high tech: Florida, Minnesota, Kansas, Colorado, and Utah. While the last three are contiguous states, their high tech loci are distinctly separated – by the Rockies and Wasatch Ranges between Utah and Colorado's "front range" cities, and by extensive plains between the latter and Wichita. The only state popularly thought of as an emerging high tech center which is missing from this set is North Carolina, which has experienced rapid high tech growth, but whose industrial economy is overwhelmingly "low tech."[3]

In some ways, this list confirms popular portrayals of Route 128 and Silicon Valley as the nation's top high tech centers. California and Massachusetts are at the hub of high tech agglomerations in their regions,

Where high tech locates

Table 7.2 High tech state agglomerations, 1977.

Regional agglomeration	High tech jobs by state	Jobs (000)	States in group with high tech location quotient greater than 1	Location quotient
Major cores				
Pacific Southwest	California	641.3	Arizona	1.80
	Arizona	45.9	California	1.49
Western Gulf	Texas	285.7	Oklahoma	1.38
	Louisiana	56.4	Texas	1.33
	Oklahoma	53.9	Louisiana	1.22
Chesapeake/	New Jersey	232.3	Maryland	1.60
Delaware River	Maryland	48.4	New Jersey	1.23
Old New England	Massachusetts	204.6	Connecticut	1.65
	Connecticut	160.0	Massachusetts	1.43
Lower Great Lakes	Illinois	360.3	Illinois	1.15
Minor cores				
	Florida	119.0	Florida	1.39
	Minnesota	91.3	Minnesota	1.18
	Kansas	62.8	Kansas	1.63
	Colorado	52.9	Colorado	1.57
	Utah	22.9	Utah	1.30

at least in terms of total numbers of jobs. However, one striking feature of the major agglomerations is that states *adjacent* to the acknowledged centers are often apt to have a higher proportion of high tech jobs in their manufacturing complexes than the *core* states. Arizona, Connecticut, and Maryland are examples. One cannot fail to be impressed, also, with the performance of states such as Florida, Minnesota, and Colorado, all of which are clearly far-distant from such traditional centers of manufacturing innovation as Chicago, New York or Los Angeles, but have nevertheless demonstrated an ability to become high tech hosts.

The major cores are, of course, not much different from a list of major manufacturing and even population centers in the USA. One reason for this is that they have been defined in part on the basis of sheer size. But it is also due to the longer-term inertia in an industrial system, whereby older manufacturing centers do give rise to new products and processes, as theories of innovation diffusion suggest. On the other hand, such core states as California and Texas are clearly more prominent in high tech sectors than they are in manufacturing in general. Furthermore, at the level of second-tier states, the list of new high tech manufacturing centers is almost entirely different from the list of basic manufacturing centers – a list which would include Ohio, Indiana, Michigan, and Missouri, the

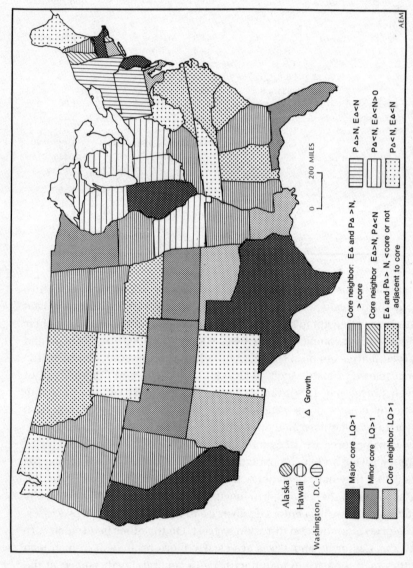

Figure 7.2 The five core areas and their fringes. LQ = location quotient, E = employment, P = population, N = national average

older centers of innovation in autos, machine tools, farm machinery, appliances and other basic capital and consumer goods. One is forced to conclude that there are indeed "new" centers of agglomeration, which did not develop out of established manufacturing activity – though they may well result from branch plants of established companies in other parts of the country.

On the other hand, one cannot hold a very rigid view that older manufacturing sectors have universally repelled new innovative sectors. Not all older manufacturing regions have lost out in the competition to become new high tech centers. Old New England is an outstanding example of a region which was until the postwar period a very intensive manufacturing center and, despite severe restructuring, has become a high tech center.[4] Illinois and New Jersey are other examples, perhaps the exceptions in the manufacturing belt, where older manufacturing complexes have managed at least through 1977 to retain high tech jobs.

The differing high tech base of the core states

What precisely is the high tech base of these five major core areas? An easy way to summarize it is to array the 100 high tech industries according to their contribution to high tech employment in each state, and to pick out those that make up half the total. When this is done (for 1977), the five cores are found to have very different high tech mixes.

The largest, the Pacific Southwest, is in many ways the clearest-cut: it is dominated by the newer post-World War II high tech industries, especially in aerospace and electronics. California, with its larger high tech base, is naturally more diverse than Arizona. Its seven leading industries are radio and TV equipment (12.0% of total employment), aircraft (9.4%), electronic computing equipment (8.7%), guided missiles (8.0%), semiconductors (5.5%), electronic components not elsewhere classified (NEC; 4.5%) and petroleum refining (2.5%). Arizona, with its later start, has a much narrower base in which only three industries contribute 50% of total high tech jobs: semiconductors (20.4%), electronic components, NEC (18.2%) and radio/TV transmitters (12.7%).

The next complex, the Western Gulf States, has a significantly different structure in which aerospace and electronics – with a clear military basis – are combined with oil extraction and chemical industries. Texas neatly illustrates this: its five leaders are oilfield machinery and equipment

(13.8%) followed by industrial organic chemicals, NEC (10.3%), aircraft (9.8%), petroleum refining (9.6%), and radio/TV transmitting equipment (6.7%). Neighboring Oklahoma's list of five industries is headed by aircraft parts (13.3%) and aircraft (12.3%) followed by oilfield machinery (10.4%), electronic computing equipment (9.0%) and construction machinery equipment (8.1%). Louisiana is even more narrowly oil–chemical–defense dominated, by just three industries: industrial organic chemicals, NEC (30.0% of the total), petroleum refining (17.2%) and guided missiles/space vehicles (6.6%).

The Chesapeake/Delaware River complex has a quite different structure, reflecting its older industrialization based on innovation in past manufacturing eras, albeit with an admixture of very new industries. Maryland illustrates this well: its five high tech leaders are radio/TV transmitting equipment with 30.6% of the total, followed by telephone and telegraphic equipment (8.1%), internal combustion engines, NEC (5.6%), power driven hand tools (3.7%) and biological products (3.4%). New Jersey shares with Illinois the most diverse high tech base of any of the core states, and the one most dominated by older industries. Eleven industries make up half its high tech job base: radio/TV transmission (9.7%), pharmaceutical preparations (9.6%), perfumes and cosmetics (6.3%), cyclic crudes, intermediates and dyes (4.3%), photographic equipment (3.9%), industrial organic chemicals, NEC (3.4%), electronic components, NEC (3.3%), electronic computing equipment (3.0%), petroleum refining (2.4%), paints, varnishes and lacquers (2.4%) and switch gear and switchboard equipment (2.3%). In Old New England, significantly, the aerospace/electronics mix quite closely resembles that of the "new" Pacific Southwest, albeit more diverse and with a mixture of some older high tech industries. Connecticut's eight leading industries are aircraft parts (20.4%) and aircraft (8.4%), followed in order by ball and roller bearings (5.9%), small arms (4.2%), radio/TV transmitting equipment (3.5%), aircraft parts and equipment (3.0%), electronic components, NEC (2.9%) and switch gear/switchboard equipment (2.9%). Massachusetts is even more diverse, with nine industries making up the 50% list: electronic computing equipment (11.7%), radio/TV transmitting equipment (10.4%), photographic equipment and supplies (8.2%), aircraft engines and parts (5.0%), electronic components, NEC (4.6%), ball and roller bearings (3.2%), industrial instruments for measuring etc. (3.1%), machine tools (2.8%) and aircraft (2.6%).

Finally, Illinois – the third high tech core within the traditional manufacturing belt – shares with New Jersey the most diverse high tech base and the one most dominated by older industries. Thirteen industries make up 50% of total high tech employment, and many of them make up a roll-call of America's industrial past: construction machinery equipment (8.9%), telephone/telegraph apparatus (7.1%), internal combustion engines, NEC (5.0%), electronic components, NEC (3.8%), radio/TV receivers (3.6%), railroad equipment (3.7%), pharmaceutical preparations (2.9%), photographic equipment (2.7%), special dies and die sets (2.6%), electronic computing equipment (2.3%), switch gear and switchboard equipment (2.3%) and paints, varnishes and lacquers (2.3%).

A clear distinction thus emerges between two archetypes: one, corresponding to the popular image, represented by Sunbelt states based on aerospace, electronics and oil-based industries; the other, represented by Frostbelt states still engaged in older high tech activities including more traditional types of engineering, telecommunications, and chemical industries. But the distinction is not completely clear-cut, because even in the traditional manufacturing belt the high tech states have a mixture of the newer industries – and, in the case of Old New England, are even dominated by them. There is a clear policy implication here: that older, traditionally based industrial states can successfully enter the high tech race. But in order for that to be true, it would be necessary for their older technologically based industries to demonstrate continued growth – which, on the basis of the analysis in Chapter 4, is unlikely. This poses the question of how, relatively, did the high tech core states perform.

Agglomeration of high tech in regional centers

Did the core states maintain their positions in the period 1972–7? The answer is a qualified yes. Ten core states are listed in Table 7.3, each of which either serves as the center of a major regional agglomeration by virtue of its sheer size or its status as a single-state minor core. Only one, New Jersey, lost high tech jobs. Another, Illinois, failed to post job growth rates in high tech greater than the national rate of high tech job creation. A third, Massachusetts, failed to host net new plants at a rate greater than the national average. The poor showing of New Jersey is clearly due to the preponderance of the older high tech sectors which are growing relatively slowly and have relatively higher dispersal rates (see Ch. 4 above). Illinois' high tech economy relies heavily upon the basic machining industries, also relatively older, slower growing and hard hit by successful import competition. These are high tech centers in which

Table 7.3 Core state high tech agglomeration, 1972–7.

	Plants			Jobs		
Core states	Share 1977 (%)	Percent 1972–7	Share of net new plants 1972–7 (%)	Share 1977 (%)	Percent 1972–7	Share of net new jobs 1972–7 (%)
California	15.5	+28.4	20.3	13.5	+17.9	25.6
Massachusetts	3.9	+12.1	2.5	4.3	+12.1	5.8
New Jersey	5.8	+7.5	2.4	4.9	−1.2	−0.7
Texas	5.3	+36.4	8.5	6.0	+20.9	12.9
Illinois	7.3	+9.8	3.9	7.6	+1.1	1.0
Florida	2.9	+31.9	4.2	2.5	+12.4	3.4
Minnesota	1.9	+30.2	3.0	1.9	+21.4	4.3
Kansas	1.1	+27.5	1.1	1.3	+10.4	1.6
Colorado	1.1	+46.3	2.0	1.1	+19.4	2.3
Utah	0.5	+50.6	0.9	0.5	+34.7	1.5
USA	100.0	+19.8	100.0	100.0	+8.7	100.0

initial agglomerative impulses have been overtaken by the slow growth and dispersion which accompanies maturation.

With these exceptions, all the other core states mounted impressive records of attracting higher shares of new jobs than they had had to begin with. California, for instance, accounted for 12.4% of all US high tech jobs in 1972, but captured 25.6% of net new jobs created nationally over the next five years. Texas, with 5.4% of national high tech employment in 1972, hosted 12.9% of net new job creation in the same period.

But if the big high tech states did well, some of the minor core states did even better at boosting their agglomerative growth. Colorado, Utah and Minnesota posted job growth rates higher than California and higher than or equivalent to Texas. While they began with much smaller bases, nevertheless they demonstrated that newer, smaller high tech centers could also garner a larger share of national high tech job growth than they had to begin with. In other words, agglomeration was still, in the mid-1970s, working in their favor.

Performance differentials do show dramatic regional cleavage although it does not fall along simple Sunbelt/Frostbelt lines. In aggregate, the New England, Middle Atlantic and east North Central regions grew more slowly than the nation in both high tech plant and job categories (see Table 7.4). And, of our five major cores, those in the Frostbelt grew distinctly less rapidly than those in the Sunbelt (Table 7.5). But to conclude that high tech agglomeration is entirely a Sunbelt phenomenon is to ignore the important evidence below the regional level. Massachusetts, for instance, a core state in a major agglomeration, hosted job growth at

Table 7.4 Plant and job change by state and region, 1972–7.

	Plants				Jobs (000)			
			Change				Change	
Region/state	1972	1977	No.	%	1972	1977	No.	%
West	8,179	10,779	+2,600	+31.8	728.3	857.0	+128.7	+17.7
Pacific	7,084	9,160	+2,076	+29.3	610.5	711.5	+101.0	+16.5
Alaska	14	16	+2	+14.3	0.2	0.5	+0.3	+150.8
California	6,251	8,029	+1,778	+28.4	544.1	641.3	+97.2	+17.9
Hawaii	39	43	+4	+10.3	0.7	0.7	+0.0	+0.2
Oregon	348	464	+116	+33.3	20.8	24.5	+3.7	+17.8
Washington	432	608	+176	+40.7	44.7	44.5	−0.2	−0.4
Mountain	1,095	1,619	+524	+47.9	117.8	145.5	+27.7	+23.5
Arizona	283	426	+143	+50.5	36.4	45.9	+9.5	+26.1
Colorado	382	559	+177	+46.3	44.3	52.9	+8.6	+19.4
Idaho	47	67	+20	+42.5	5.0	7.6	+2.6	+52.0
Montana	41	53	+12	+29.2	1.5	2.2	+0.7	+46.7
Nevada	49	90	+41	+83.7	1.8	3.5	+1.7	+94.4
New Mexico	101	139	+38	+37.6	9.5	8.3	−1.2	−12.6
Utah	160	241	+81	+50.6	17.0	20.9	+5.9	+34.7
Wyoming	32	44	+12	+72.7	2.3	2.2	−0.1	−4.3
South	8,028	10,525	+2,497	+31.1	991.0	1,153.6	+62.6	+16.4
West South Central	3,011	4,080	+1,069	+35.5	353.2	434.0	+80.8	+22.9
Arkansas	187	268	+81	+43.3	25.7	38.0	+12.3	+47.9
Louisiana	365	456	+91	+24.9	51.0	56.4	+5.4	+10.6
Oklahoma	414	567	+153	+37.0	40.1	53.9	+13.8	+34.4
Texas	2,045	2,789	+744	+36.4	236.4	285.7	+49.3	+20.9

Table 7.4 Plant and job change by state and region, 1972-7 – continued.

	Plants				Jobs (000)			
			Change				Change	
Region/state	1972	1977	No.	%	1972	1977	No.	%
East South Central	1,401	1,805	+404	+28.8	190.3	211.9	+21.6	+11.4
Alabama	334	430	+96	+28.7	44.6	48.9	+4.3	+9.6
Kentucky	333	420	+87	+26.1	42.9	49.0	+6.1	+14.2
Mississippi	176	224	+48	+27.3	21.7	26.0	+4.3	+19.8
Tennessee	558	731	+173	+31.0	81.1	88.0	+6.9	+8.5
South Atlantic	3,616	4,663	+1,047	+29.0	447.5	507.7	+60.2	+13.5
Delaware	86	87	+1	+1.2	11.4	10.4	−1.0	−0.1
Florida	1,155	1,524	+369	+31.9	105.9	119.0	+13.1	+12.4
Georgia	554	700	+146	+26.4	41.9	51.0	+9.1	+21.7
Maryland	451	529	+78	+17.3	52.4	48.4	−4.0	−7.6
North Carolina	528	693	+165	+31.2	72.4	93.0	+20.6	+22.1
South Carolina	244	361	+117	+48.0	55.4	71.0	+15.6	+28.2
Virginia	387	524	+137	+35.4	74.6	81.7	+7.1	+9.5
West Virginia	183	222	+39	+21.3	31.3	31.7	+0.4	+1.3
Washington DC	28	23	−5	−17.9	1.2	0.5	−0.7	−58.3
Northeast	13,065	14,051	+986	+7.5	1,301.5	1,322.5	+21.0	+1.5
Middle Atlantic	9,337	9,755	+418	+4.5	899.1	883.4	−15.7	−1.8
New Jersey	2,849	3,063	+214	+7.5	235.1	232.3	−2.8	−1.2
New York	3,981	3,972	−9	−0.2	347.0	336.8	−10.2	−2.9
Pennsylvania	2,507	2,720	+213	+8.5	317.0	314.3	−2.7	−0.9
New England	3,728	4,296	+568	+15.2	402.4	439.1	+36.7	+9.1
Connecticut	1,300	1,482	+182	+14.0	157.3	160.0	+2.7	+1.7
Maine	89	118	+29	+32.6	11.3	9.0	−2.3	−20.4
Massachusetts	1,826	2,047	+221	+12.1	182.4	204.6	+22.2	+12.1

New Hampshire	187	268	+81	+43.3	22.2	29.5	+4.3	+32.9
Rhode Island	258	304	+46	+17.8	16.4	18.8	+2.4	+14.6
Vermont	68	77	+9	+13.2	12.8	17.2	+4.4	+34.4
North Central	14,875	16,715	+1,840	+12.4	1,358.8	1,426.7	+67.9	+5.0
West North Central	2,599	3,189	+590	+22.7	277.8	319.7	+41.9	+15.1
Iowa	354	467	+113	+31.9	41.2	54.0	+12.8	+31.1
Kansas	360	459	+99	+27.5	56.9	62.8	+5.9	+10.4
Minnesota	771	1,004	+233	+30.2	75.2	91.3	+16.1	+21.4
Missouri	889	962	+73	+8.2	82.9	87.0	+4.1	+4.9
Nebraska	159	198	+39	+24.5	17.1	18.7	+1.6	+9.4
North Dakota	25	42	+17	+68.0	1.3	1.9	+0.6	+46.2
South Dakota	41	57	+16	+39.0	3.2	4.0	+0.8	+25.0
East North Central	12,276	13,526	+1,250	+10.2	1,081.0	1,107.0	+26.0	+2.4
Illinois	3,506	3,849	+343	+9.8	356.5	360.3	+3.8	+1.1
Indiana	1,194	1,346	+152	+12.7	146.7	153.4	+6.7	+4.6
Michigan	3,319	3,619	+300	+9.0	159.0	169.4	+10.4	+6.5
Ohio	3,127	3,409	+282	+9.0	293.7	295.1	+1.4	+0.5
Wisconsin	1,130	1,303	+173	+15.3	125.1	128.8	+3.7	+3.0
Total	44,147	52,903	+8,756	+19.8	4,379.8	4,759.8	+380.0	+8.7

Table 7.5 High tech state agglomerations: plant and employment change, 1972–7.

	Plant change		Employment change (000)	
	No.	%	No.	%
Arizona	143	50.5	9.5	26.1
California	1,778	28.4	97.2	17.9
Pacific Southwest	1,921	29.4	106.7	18.4
Oklahoma	153	37.0	13.8	34.4
Texas	744	36.4	49.3	20.9
Louisiana	91	24.9	5.4	10.6
Western Gulf	988	34.9	68.5	20.9
Maryland	78	17.3	−4.0	−7.6
New Jersey	214	7.5	−2.8	−1.2
Chesapeake/ Delaware River	292	8.8	−6.8	−2.4
Connecticut	182	14.0	2.7	1.7
Massachusetts	221	12.1	22.2	12.1
Old New England	403	12.9	24.9	7.3
Illinois	343	9.8	3.8	1.1
Lower Great Lakes	343	9.8	3.8	1.1

greater than the national pace. So did Minnesota, an important minor core, which is one of the frostiest places in the country. What does seem to be clear is that high tech plant location and job growth is shunning the older midwest industrial belt from Buffalo to St. Louis and Milwaukee (with the sole exception of Chicago), a pattern which may be connected to the emerging defense perimeter (see Markusen 1985b).

The regional dispersion of high tech industries within major agglomerations

While most core states did continue to attract a disproportionate share of high tech jobs in the mid-1970s, they were frequently outpaced by their nearest neighbors, indicating that dispersion had begun to take place within agglomerative regions. Higher plant and job growth rates were typically experienced by states contiguous to the core state, while states on the outer ring of the region experienced slower growth. This distinctive pattern is consistent with our findings in Chapter 6 that most high tech sectors are dispersing geographically.

This pattern is perhaps best illustrated by the events in the westernmost region – that area with California as its core and all the states from Idaho, Nevada and New Mexico westward as its ring (see Fig. 7.2). California, which accounted for more than 25% of all net new jobs and 20% of plants created in the period, was equalled or surpassed in its job growth rate by the neighboring states of Arizona and Nevada. Of these two, only Arizona was more specialized in high tech manufacturing than California and the nation. It is important to keep in mind that California accounted for 75% of all high tech employment in the entire Pacific and Mountain regions in 1977, so that these neighbors are small employers by comparison; the three states together accounted for just 74,000 high tech jobs in 1977. Yet the extraordinary growth rates of this inner ring do show that dispersion from the core had started by the early 1970s.

One point must be stressed about this process: the figures do not tell us anything about the actual mechanisms. Simply because high tech grows faster in Arizona than in California, it is not necessarily the case that firms are moving from California to Arizona or that California firms are setting up branch plants there; it could equally be the case that the new activity is spawned by firms from New England or elsewhere. In fact, we know that some of the Arizona plants are owned by companies based in the Mid-west (Glasmeier 1986).

States yet further removed from California's sphere of influence had even more erratic growth patterns and were of relatively minor importance. Alaska added two plants, which boosted that state's high tech employment by 150%. On the other hand, Hawaii fell below the rest of the region, and the nation, in both job and plant additions.

This western pattern is typical of the rest of the country, with the exception perhaps of the Middle Atlantic region. In every major region, and in several of the minor ones, at least one state immediately surrounding the high tech core experienced faster high tech growth rates than the core. Most of these were not states that themselves specialized in high tech production. In the Western Gulf region, Texas' and Louisiana's neighbors, Oklahoma and Arkansas, hosted new plant and job growth in excess of that core, although it is important to keep in mind that some of this was oil-crisis related and that overall levels were not high – 25,000 high tech jobs in total in 1977. Mississippi also posted an impressive growth rate from a small base. In New England, Vermont, and New Hampshire gained high tech activity at rates above Massachusetts', but Connecticut and Maine, although experiencing higher plant growth, lagged behind in net job creation.

In the older manufacturing belt, the dispersion experience was mixed. The states around Illinois did fit the general pattern, with Wisconsin, Indiana, Michigan, and Kentucky all posting high tech job growth rates in excess of Illinois. Michigan's relatively high rate of net high tech job creation may be an indigenous phenomenon, however, associated with dramatic retooling in the auto industry.

The one striking exception to the national pattern occurred in the Chesapeake/Delaware River region. Here New Jersey's neighbors, with the exception of Pennsylvania, lost high tech employment faster than did the core state itself. This suggests that here dispersion of high tech within the region is not prevalent. Rather, two forces seem to be at work in this area. One, dispersion, may be occurring through interregional or even international rather than intraregional relocation and movement. The other consists of differences in product mix whereby substitute and superior commodities, developing in newer regions, crowd out this older region's specialties – a process not of relocation but of displacement, though in a more indirect fashion. But this explanation in itself cannot account for the faster rates of high tech loss in the neighboring states than in the core state. The process here may be that during the profit-squeeze period of rationalization, cutbacks, and restructuring, the branch plants which were spun off in an earlier period are closed before main plants which have remained the experimental showpieces of the firm's operations. This, however, is speculative.[5]

The larger of the minor core states seem also to have spun off high tech plants and jobs to surrounding states, although their spheres of influence do not appear to have reached past the immediately bordering states. Minnesota's neighbors, the two Dakotas and Iowa, all posted growth rates in excess of that state's performance, though again the high tech total – 60,000 jobs in 1977 for the three states together – was not all that high. Georgia's job growth rate surpassed Florida's, although its plant growth rate was slightly below hers; Alabama lagged Florida but North Carolina, not identified as a core high tech agglomeration by our criterion, nevertheless deserves mention as a state which not only posted high job growth rates in the mid-1970s, but also seems to have abetted even higher job growth rates in South Carolina.

Only the smallest of the high tech core states – the three rather remote centers of Utah, Colorado, and Kansas – failed to generate high tech spin-offs of notable magnitude in neighboring states.[6] High tech migration from California accounts for Idaho's strong showing, but in total this state's growth took place in just 20 plants and created only 2,600 new jobs. Wyoming had 12 more plants at the end of the period than at the outset, but actually lost employment. (Nebraska exceeded national growth rates but lagged behind its high tech neighbors.) Yet Utah, Colorado, and Kansas remain strong contenders in the high tech race; together they accounted for almost 140,000 high tech jobs in 1977, had increased their shares in the 1970s, and all showed extraordinarily high levels of high tech specialization.

One might infer that a sort of "optimal high tech complex size" may have been operating during this period. All core states with a high tech workforce in excess of 70,000 seem to be associated with significant growth in at least some surrounding states. Those with fewer than this many employees did not, but they still led their neighbors in rates of high tech growth. This generalization must be viewed as temporally circumscribed and highly colored by the artifacts of state boundaries.

A final note is that there do appear to be border states between high tech agglomerations or on the periphery which have not enjoyed high tech spinoffs. The states of Maine, Ohio, West Virginia, Tennessee, New Mexico, Washington, and Hawaii are all examples of poor performance in high tech job growth. Ironically, some of these states have long been among the nation's major manufacturing states. Evidently, neither a history of "low tech" manufacturing growth nor past hosting of innovative industries like machine tools (Ohio) and aerospace (Washington) guarantee that a state will be in on the current high tech wave. This underscores our view that high tech is a distinctly different phenomenon from manufacturing in general.

Metropolitan high tech agglomerations

High tech agglomerations can be more precisely studied at the metropolitan level. When viewed close up, metropolitan areas demonstrate powerfully the tendency for high tech activity to disperse from older core areas to adjacent metropolitan units. In this final section, we look at the evidence for agglomerative and dispersing tendencies at the metropolitan level.

Top metropolitan rankings, by job and plant levels and growth performance

Within state groupings, some metropolitan areas, sometimes contiguous, account for a disproportionate share of jobs.[7] A group of the nation's

Table 7.6 Top ranked metropolitan areas, job and plant levels and change, 1972–7.

Rank	SMSA	Plants, 1977
1	Los Angeles–Long Beach, California	3,732
2	Chicago, Illinois	3,029
3	Detroit, Michigan	2,291
4	New York, New York/New Jersey	2,149
5	Boston–Lowell–Brockton–Lawrence–Haverhill, Massachusetts	1,484
6	Philadelphia, Pennsylvania/New Jersey	1,455
7	Anaheim–Santa Ana–Garden Grove, California	1,118
8	Newark, New Jersey	1,077
9	Cleveland, Ohio	1,037
10	Nassau–Suffolk, New York	963
11	Dallas–Fort Worth, Texas	942
12	San Francisco–Oakland, California	933
13	San Jose, California	856
14	Houston, Texas	840
15	Minneapolis–St. Paul, Minnesota/Wisconsin	740

Rank	SMSA	Jobs, 1977
1	Los Angeles–Long Beach, California	279,293
2	Chicago, Illinois	255,051
3	Boston–Lowell–Brockton–Lawrence–Haverhill, Massachusetts	144,720
4	Philadelphia, Pennsylvania/New Jersey	136,891
5	San Jose, California	106,002
6	Anaheim–Santa Ana–Garden Grove, California	92,726
7	Newark, New Jersey	92,078
8	Dallas–Fort Worth, Texas	87,658
9	Detroit, Michigan	87,180
10	Houston, Texas	81,577
11	New York, New York/New Jersey	80,980
12	Milwaukee, Wisconsin	69,741
13	Minneapolis–St. Paul, Minnesota/Wisconsin	68,664
14	Cleveland, Ohio	66,344
15	Nassau–Suffolk, New York	66,335

Rank	SMSA	Net plant change 1972–7
1	Anaheim, California	464
2	Los Angeles, California	367
3	San Jose, California	339
4	Dallas, Texas	276
5	Chicago, Illinois	224
6	Houston, Texas	204
7	Boston, Massachusetts	191
8	Minneapolis, Minnesota	158
9	San Francisco, California	151

Table 7.6 Top ranked metropolitan areas, job and plant levels and change, 1972–7 – *continued*.

Rank	SMSA	Net plant change 1972–7
10	Detroit, Michigan	145
	Median gain	9

Rank	SMSA	Net job change 1972–7
1	San Jose, California	31,909
2	Anaheim, California	30,612
3	Houston, Texas	18,932
4	San Diego, California	16,782
5	Boston, Massachusetts	15,173
6	Dallas, Texas	12,067
7	Worcester, Massachusetts	9,893
8	Oklahoma City, Oklahoma	8,363
9	Lakeland, Florida	8,132
10	Phoenix, Arizona	7,976
	Median gain	248

Rank	SMSA	Percent plant change
1	Lawton, Oklahoma	600.00
2	St. Cloud, Minnesota	214.29
3	Laredo, Texas	150.00
4	Santa Cruz, California	137.50
5	Champagne–Urbana, Illinois	118.18
6	Oxnard, California	114.55
7	Fort Meyers, Florida	110.00
8	Billings, Montana	100.00
9	Cedar Rapids, Iowa	100.00
10	Panama City, Florida	100.00

Rank	SMSA	Percent job change
1	Lawton, Oklahoma	2,266.97
2	St. Cloud, Minnesota	1,265.08
3	Boise, Idaho	729.31
4	Santa Rosa, California	360.58
5	Lakeland, Florida	266.69
6	Lubbock, Texas	237.44
7	Topeka, Kansas	237.35
8	Laredo, Texas	220.84
9	Savannah, Georgia	204.78
10	McAllen–Pharr–Edinburg, Texas	181.54

largest metropolitan areas dominate the top ten rankings for total jobs and plants in 1977 (see Table 7.6, sections 1 & 2). In sheer numbers of plants, northeastern SMSAs still predominate. In job totals, the top ten are evenly split between Frostbelt and Sunbelt locations. Several metropolitan areas – Boston, San Jose, and Anaheim – rank better than they would in an array by sheer population or workforce size. New York is conspicuously absent from the job column, while its neighbor Newark makes the list. For both totals, the top ten include four metropolitan areas – Anaheim, Newark, Nassau–Suffolk, and San Jose – "adjacent" to or contiguous with older, commercial cities which by 1977 accounted for greater numbers of plants and/or jobs than their parent (Glasmeier 1986).

Relative job and plant shifts during the mid-1970s heavily favored big metropolitan areas outside the traditional manufacturing belt (see sections 3 and 4 of Table 7.6). Only Chicago, Boston and Detroit enjoyed net plant additions substantial enough to make the top ten, while Boston and Worcester SMSAs were the only northeastern SMSAs to make the jobs rankings. The top job generating metropolitan areas are San Jose, Anaheim, Houston, San Diego, Boston, Dallas, and Worcester, in that order. Surprising performances in net job gain were posted by Oklahoma City, Lakeland (Florida), and Phoenix. Top ratings in the percentage change category (sections 5 and 6) went to small metros whose job and plant gains looked impressive on the top of a minute pre-existing base.[8]

One intriguing commonality among this set is that they are almost all "interior" sites and only very few of them are contiguous with larger metropolitan high tech centers. This set may be representative of those smaller metros who are receiving branch plants and spinoffs of the more routine parts of high tech production. With few exceptions, they are not places with important high tech agglomerative inducements such as major universities (except Champagne–Urbana), well-developed business service sectors, or the presence of Fortune 500 headquarters (Boise and Savannah are exceptions in this last category).

The rankings just presented reflect the way in which SMSA boundaries have evolved. Somewhat artificially, then, the largest urban concentrations – such as those in the Los Angeles basin and those in the greater Bay Area – are treated as a group of separate SMSAs, while free-standing SMSAs, such as Houston and Dallas, appear more significant than they would if high tech agglomerations were aggregated in more natural units. It is thus preferable to look more carefully at larger substate agglomerations of high tech activity, which is what we do in the rest of this chapter.

Change within metropolitan agglomerations

The basic pattern of high tech location and change within metropolitan agglomerations is roughly the following. An older, commercial era core SMSA still, in 1977, contained the greatest number of high tech plants, and in some cases, jobs as well. One or more neighboring SMSAs accounted for the greatest number of net new jobs created in the mid-1970s, and sometimes served as the true center of high tech agglomeration by accounting for more jobs in total than the older anchor SMSA. Outlying SMSAs, growing rapidly in the 1970s, accounted for relatively small numbers of net new plants and jobs but posted the greatest percentage gains. However, in all cases, movement of the center of high tech activity toward metropolitan perimeters was a very uneven process, accelerating in some directions and proceeding very slowly in others. Indeed, in a large number of cases, contiguous SMSAs actually lost jobs and plants in the period, even in the most prominent Sunbelt high tech complexes.

Not all growth in adjacent SMSAs represents relocation or spinoff job generation from the core; some, indeed, represents interregional shifting. As new seedbeds of innovation create around them the professional/technical labor pools and business service complexes necessary for high tech commercialization, they draw branch plants and relocations from metropolitan areas in other parts of the country (Glasmeier 1986b). But the predominant pattern, supported by evidence from secondary sources, is from innovative centers outward.

The Los Angeles Basin The Los Angeles Basin neatly demonstrates this pattern of outward movement. When viewed in terms of absolute numbers of jobs, the largest high tech agglomeration in the USA is the greater Los Angeles area (see Fig. 7.3). Including San Diego, the basin accounted for over 450,000 jobs in 1977, almost 100,000 more than the greater New York area, the second largest agglomeration. The Los Angeles SMSA, the core of the region and the nation's top metropolitan area in terms of both absolute jobs and plants in 1977, was nevertheless one of the poorest growth performers in the mid-1970s. Its high tech employment base declined by 3,220, or 1%; while this may reflect a few large defense-related plant closings, it is notable that not enough new high tech jobs were created in other sectors to override this effect.

What is striking in the Los Angeles area is the extraordinary growth rate of the Anaheim–Santa Ana–Garden Grove SMSA, sandwiched

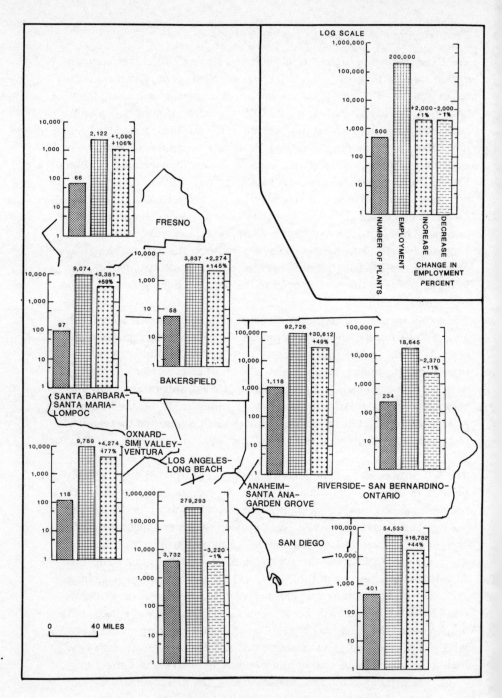

Figure 7.3 Change in plants and jobs, Greater Los Angeles, 1972–7.

between Los Angeles and San Diego. This adjacent SMSA gained over 30,000 jobs in the mid-1970s, placing it at the top of the plant growth rankings and second on job growth nationally. Dispersion from the Los Angeles core was also clearly moving up the coast and spilling over the mountains into the lower end of the Central Valley. The adjacent SMSAs of Oxnard–Ventura (77% increase), Santa Barbara (59%), Bakersfield (145%), and Fresno (106%) all topped the Anaheim (49%) and San Diego (44%) growth rates. These four northerly counties accounted for over 11,000 net new jobs in high tech manufacturing in the five years studied. One adjacent area shared in Los Angeles' net job decline in the period: Riverside–San Bernardino–Ontario, containing a relatively mature manufacturing complex, which is highly defense-related, and which may have also suffered from the defense build-down.

Chesapeake to Hudson Similar patterns of uneven rates of dispersion can be observed within the region which spans the upper Chesapeake Bay to the Hudson River (Fig. 7.4). Here, an older commercial center, New York, still accounts for the largest number of high tech plants but for fewer jobs than its neighbor, Newark. While New York lost almost 9,000 jobs in the period, the greatest job gains in the area were posted by the Nassau–Suffolk SMSA (5,744), the Newark SMSA (2,294) and the Long Branch–Asbury Park SMSA (2,810). Two neighboring SMSAs also experienced negative job growth (Peterson–Clifton–Passaic and Jersey City), while the Bridgeport, New Brunswick and Newark SMSAs all grew modestly, though by less than the national norm. Indeed, only the Long Island and northeastern shore areas of New Jersey added high tech employment at better than the national rate of 8.7%.

Another distinct high tech agglomeration stretches from Trenton to Baltimore. All five of the SMSAs in this corridor suffered heavy net losses in high tech employment in the mid-1970s, except Wilmington which grew at a rate below 1%. This is a region with heavy concentrations in those high tech sectors identified in Chapter 4 as mature and/or declining, especially in the chemicals area. No apparent centrifugal pattern characterized these declines.

The Greater San Francisco Bay Area The Greater San Francisco Bay Area demonstrates well the tendency for growth to spin outward in selective directions (Fig. 7.5). The core San Francisco–Oakland SMSA still accounted for more high tech plants in 1977 than did the Silicon Valley SMSA (San Jose), but the latter had more than twice as many jobs. While

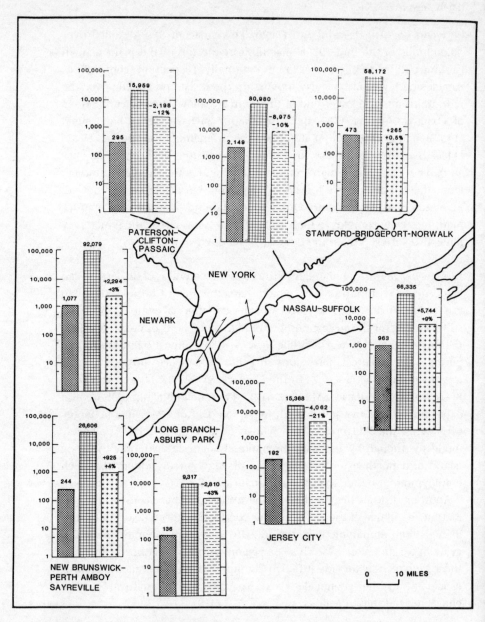

Figure 7.4 Change in plants and jobs, Chesapeake to Hudson, 1972–7. For key see Figure 7.3.

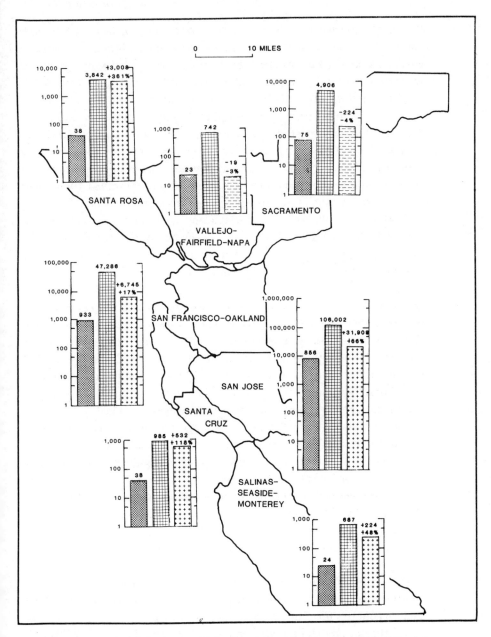

Figure 7.5 Change in plants and jobs, Greater San Francisco Bay, 1972–7. For key see Figure 7.3.

the former posted an impressive 17% job growth rate in the period, almost twice the national rate, San Jose added almost four times as many jobs as its older neighbor. High tech growth has spilled over the waterways and coast range of the highly constrained inner Bay Area. Santa Cruz boosted its job totals by 118% from a small base, a trend that has accelerated since 1977, and peripheral SMSAs of Santa Rosa and Salinas–Monterey posted dramatic gains as well. Santa Cruz is an outstanding example of spillover; engineers and entrepreneurs living in Santa Cruz but working in Silicon Valley have been the major factor in start-ups and relocations closer to home (Gordon & Kimball 1986). High tech prosperity did not radiate out in a northeasterly direction, however; the SMSAs of Vallejo–Fairfield–Napa and Sacramento actually posted modest high tech job losses in the same period. In the period since 1977, these areas appear to have turned around as well and received a greater share of area-wide high tech development.

Old New England The Boston area demonstrates a variation on the dispersion theme (Fig. 7.6). The Boston NECMA (New England Consolidated Metropolitan Area), an amalgam of SMSAs, not only accounted for the bulk of jobs in high tech in the larger eastern New England area, but accounted for the majority of new jobs created in the period as well. Its neighbors, the Worcester–Leominster area (with its Lincoln Laboratory) and the Providence SMSA added new high tech jobs at a faster rate, while its southern neighbor, Fall River–New Bedford, was a net high tech loser. Within the Boston SMSA, high tech growth has heavily concentrated in the outer rings. The entrepreneurial activity of engineers teaching in and graduating from the area's major universities has been a principal growth factor. The relatively strong performance of the core SMSA in this case is also a tribute to the area's efforts to revitalize a longer-term stagnant region and to convert under-utilized industrial space, such as in the old mill town of Lowell, to new forms of innovative manufacturing activities.

Lower Lake Michigan The SMSAs clustered around Chicago offer yet another study in intraregional dispersion (Fig. 7.7). The Chicago SMSA itself accounted for more than 255,000 jobs in 1977, ranking it number two in the nation in both high tech job and plant rankings. It grew rather slowly in the period, however, by 1,880 jobs or 1%, and within the area dramatic shifts in manufacturing location occurred from the central city of Chicago toward its suburbs (McDonald 1984). The four SMSA's surrounding Chicago all added high tech jobs at a rate exceeding the core

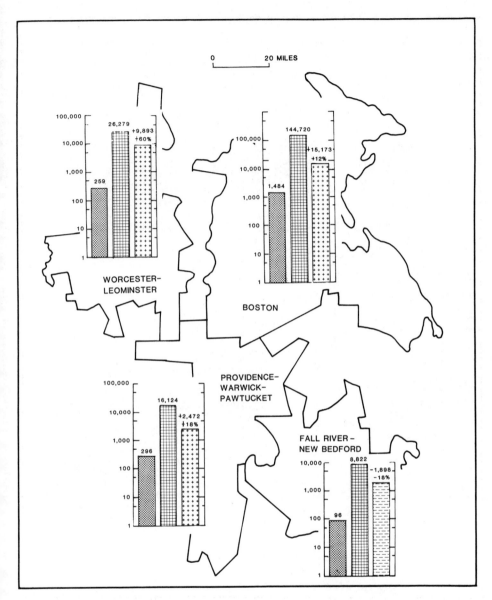

Figure 7.6 Change in plants and jobs, Old New England, 1972–7. For key see Figure 7.3.

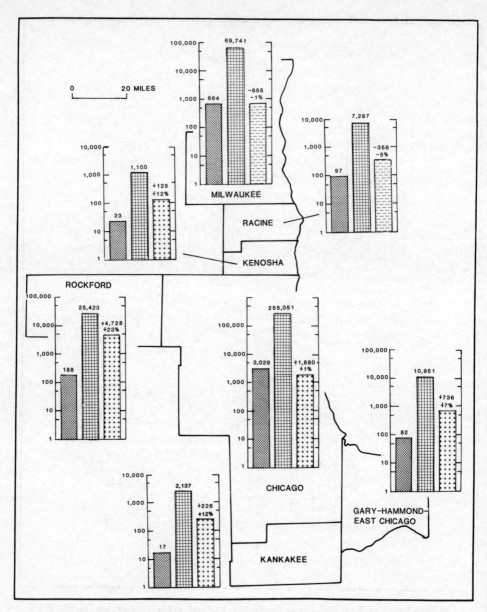

Figure 7.7 Change in plants and jobs, lower Lake Michigan, 1972–7. For key see Figure 7.3.

SMSA, and three of these, Rockford, Kenosha, and Kankakee, added them at rates exceeding the national average. Chicago presents an interesting contrast to its regional partner, Milwaukee. The latter, in itself a large high tech employer (almost 70,000 jobs in 1977), experienced net job loss in the mid-1970s, as did its Chicago-side neighbor, Racine. The relatively poorer performance of the Milwaukee/Chicago complex nationally is due to the area's heavy reliance on the machining and machine tools portions of high tech manufacturing, sectors of maturation and change where imports from Europe and Japan have made serious inroads since the late 1960s (Great Lakes Commission 1984).

The Florida high tech complexes In Florida, two quite distinct high tech complexes have evolved, with the state's other metropolitan areas by and large losing out completely in the competition for high tech, at least through 1977. The bulk of Florida's new high tech growth has occurred in the agglomeration which stretches from Tampa and St. Petersburg on the Gulf Coast to Melbourne and Titusville on the eastern coast (Fig. 7.8). Orlando can be considered its core, but high tech employment is fairly evenly distributed across the five contiguous SMSAs. Growth rates have been higher in all the surrounding SMSAs than in Orlando. The performance of these outlying SMSAs, however, is largely explained by unique factors such as the location of Cape Canaveral on the eastern coast and the presence of a number of important military facilities in the area.

An older high tech agglomeration, in the Miami/Fort Lauderdale area, was actually a net job loser in the 1970s. The Fort Lauderdale SMSA accounted for the largest number of high tech jobs among the three contiguous SMSAs, and actually grew by 70% in the period. But the two straddling areas, West Palm Beach and Miami, both lost jobs in rather large numbers. Again, since military-related high tech is prominent in these areas, especially the older aircraft industry, this could be a phenomenon related to defense procurement. If so, it presents a striking contrast to the space program related growth farther north in the state.

The Colorado front range A post-World War II, emerging high tech agglomeration is developing in the strip of cities along the front range of the Rockies, centered in Denver but spreading as far north as Fort Collins and as far south as Colorado Springs and perhaps Pueblo (Fig. 7.9). The Denver–Boulder SMSA, which came into prominence in the race-to-space program of the 1960s, accounts for the majority of plants and jobs, and

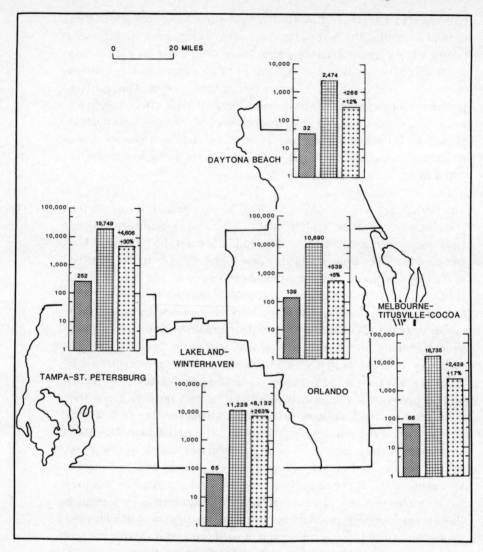

Figure 7.8 Change in plants and jobs, Florida, 1972-7. For key see Figure 7.3

grew more than twice as fast as the nation during the 1970s. The neighboring SMSAs to the north – Fort Collins and Greeley – grew even faster than Denver during the 1970s. Since 1977, and especially in the early 1980s, Colorado Springs has also taken off as a high tech development center. Much of the more recent development is military-related. Pueblo, an older coal mining and steelmaking center, has for the most part not shared in the high tech boom.

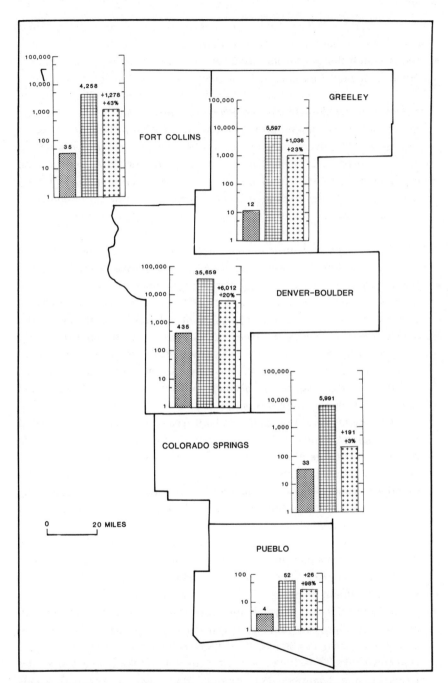

Figure 7.9 Change in plants and jobs, the Colorado front range, 1972–7. For key see Figure 7.3.

The Salt Lake corridor Utah's population is heavily concentrated in a corridor stretching from Great Salt Lake in the north southward along the Wasatch Ranges. A defense-oriented manufacturing establishment, much of it high tech, has been a strong stimulant to postwar growth (Arrington & Jensen 1965). There are two contiguous SMSAs in the corridor: Salt Lake–Ogden and Provo–Orem (Fig. 7.10). The Salt Lake metropolitan area added an impressive 4,871 jobs in the mid-1970s, an increase of 33% over 1972, and its southern neighbor added more than 500, an increase of 143%. This area can be considered an outstanding example of the catalytic role of defense facilities in stimulating high tech development. Utah's high tech complex represents not dispersion from other areas, but an indigenous, defense-nurtured phenomenon.

Stand-alone high tech metropolitan agglomerations The above profiles have been confined to cases where the Census Bureau has defined contiguous areas as separate SMSAs, permitting comparisons of intraregional dispersion. Several SMSAs have been left out of this analysis, because they "stand alone" – often because their youthfulness has permitted annexation of surrounding areas before they could become politically independent, and in some cases because county units, the basis of all SMSAs, are relatively immense. Since several states with high locational quotients have not figured in this metropolitan discussion, a brief look at their metropolitan complexes is in order.

In Minnesota, the Minneapolis–St. Paul area accounted for an estimated 68,665 high tech jobs in 1977 and grew in the mid-1970s at above the national rate (plants) and just below (jobs). The twin cities do seem to have generated growth in the neighboring but not contiguous SMSAs of St. Cloud and Duluth, both of which added jobs in the 1970s at above the national rate. In Kansas, the bulk of new job generation was concentrated in the detached SMSAs of Topeka and Wichita. Kansas City, both its Kansas and Missouri portions, actually lost over 1,000 jobs, a rate of decline of 4%. Wichita by itself accounted for more than 36,000 jobs in 1977 and added over 2,000 more in the period, a growth largely explained by the prominent presence of aircraft manufacture.

The states of Texas and Arizona each host two or more metropolitan areas with impressive high tech growth patterns. Dallas–Fort Worth added more than 12,000 jobs to its high tech complex in the 1970s, and Houston added nearly 19,000. Houston's two older neighbors, the Galveston and Beaumont–Port Arthur SMSAs, posted very modest jobs gains connected with their shipping, shipbuilding and oil-related manu-

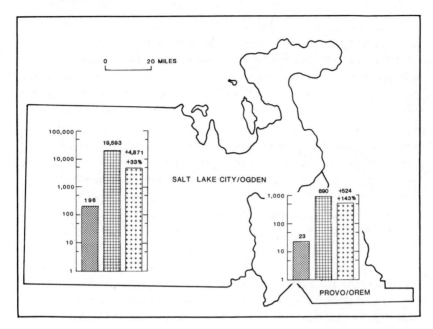

Figure 7.10 Change in plants and jobs, the Salt Lake corridor, 1972–7. For key see Figure 7.3.

facturing complexes; more recently these areas have suffered severe post-energy crisis job losses (Weinstein et al. 1985). Austin, one of the most frequently mentioned high tech centers in the 1980s, was already booming in the mid-1970s – mainly due to the establishment of branch plants – posting a high tech job growth rate of 75%. Much larger than Austin, Phoenix posted job growth gains of 28% while the smaller Tucson grew just below the national average at 8%. But while the establishment of the MCC (Microelectronics and Computer Technical Corporation) research center offered the possibility of turning Austin into a true high tech innovative center (Malecki 1986), Phoenix's high tech sector continued to be dominated by branch plants (Glasmeier 1986).

Summary

The evidence, then, at the state and metropolitan levels does indicate that dispersion is taking place within and from major high tech agglomerations. The distribution of high tech manufacturing employment is increasingly distinct from that of older, heavy industry in the United

States. It has historically been well represented in the older manufacturing belt spanning Milwaukee to St. Louis to Baltimore to New York, but it has in recent years suffered setbacks in the more mature sectors, setbacks which have not been counterbalanced by a complementary share of newer innovative high tech sectoral growth. Deindustrialized New England, Long Island, and Minnesota are the exceptional, and peripheral, areas of the older manufacturing belt which have successfully bucked the tide of dispersion and managed to garner a share of newer high tech sectoral agglomeration. They ignore the Sunbelt/Frostbelt divide. Within the older manufacturing belt, some outlying SMSAs have also benefited from intraregional dispersion, but their experience is not generally replicated by most other peripheral areas and medium-sized detached SMSAs in the region.

Conversely, it is a mistake to equate high tech concentration with the Sunbelt. That term, in so far as it has a conventional meaning, embraces the whole vast arc from Virginia and the Carolinas in the east, through Georgia and the Gulf States, to such mountain and desert states as New Mexico and Arizona, and finally to California. In most of the states in this arc, high tech is in fact not well represented. Even North Carolina, which in popular repute is associated with the Research Triangle, shows up relatively poorly on our measures. Sunbelt high tech, in fact, is concentrated in a very few states: Florida, part of Texas (especially the south), the mountain and desert states, and California.

While we have not tested explicitly for the relative contribution of each to the pattern of dispersion, it does seem clear that two quite distinct forces are at work overall. First, new centers of agglomeration in high tech industry have evolved, and – with important exceptions, like Boston – their cores are not coterminant with older centers of innovation in manufacturing. In other words, the newer sectors have different centers of gravity from the older ones. We suspect that this is the major factor behind the interregional dispersion observed.[9] At the same time, dispersion of the more standardized high tech activities within sectors has taken place, both toward urban peripheries, toward smaller cities within the same regions, and even interregionally. This has been truer of the more mature high tech sectors, and thus is a stronger factor in explaining high tech performance in older and more easterly metropolitan areas than it is in the south and west. Nevertheless, this source of dispersion is also quite marked within the faster-growing agglomerations like the Los Angeles Basin, the Bay Area and smaller complexes like Colorado and Utah's mid-state corridor. One explanation is that innovative high tech sectors spin off certain functions and search for lower-cost production sites.

What is also clear from the descriptive analysis in this chapter is that not all peripheral places can expect the process of dispersion to provide painless, automatic plant and job growth for their workforces. In almost every metropolitan agglomeration studied, some large high tech sectors experienced declines. Nor did all same-sized metropolitan areas in the Sunbelt experience similar gains. The question must thus be faced, "What factors determine the ability of a metropolitan area to generate and/or attract plant and job growth within its borders?" To that question, *why* high tech locates and expands where it does, we now turn.

Notes

1. An example would be the separation of Maryland and Delaware from New Jersey, Minnesota from Wisconsin, California from Nevada.
2. The location quotient is computed as follows: $(HT_r/HT_t)/(M_r/M_t)$ where HT_r = high tech employment in state r, HT_t = national high tech employment, M_r = total state manufacturing employment and M_t = total national manufacturing employment. A location quotient of more than 1 indicates a specialization in high tech relative to the nation as a whole (Glasmeier 1985).
3. North Carolina ranked 43rd of the 50 states in share of manufacturing accounted for by high tech. See Glasmeier (1986b), Table 5.
4. However, the relatively high location quotient here is as much a function of the *loss* of textile, shoe and basic machinery jobs as it is the growth of high tech. See Harrison (1982).
5. A theoretical interpretation of this potential for reconcentration, and some tentative evidence for it, can be found in Markusen (1985a), Ch. 5.
6. States like Oklahoma, Arkansas, and Arizona clearly belong within the spheres of influence of other high tech cores.
7. Nationally, metropolitan areas accounted for over 80% of high tech employment in 1977. Non-SMSA high tech was heavily dominated by chemical- and petroleum-related activities (Markusen, Hall and Glasmeier 1984).
8. It should be kept in mind that the job estimates for individual metropolitan areas could be biased, especially if one of the plants they hosted was in the open-ended 1,000+ employment interval. See the discussion in Chapter 5.
9. Our findings provide suggestive evidence for the conclusions of Norton and Rees (1979) in their product cycle study of metropolitan shifts. They also argued that new seedbeds of innovation were springing up in the South and West, showing this with a shift-share analysis.

8

Why high tech locates

Toward a theory of location

IN CHAPTERS 3–6, we explored the phenomenon of high tech geographical distribution from the point of view of the industries themselves. In Chapter 7, we shifted to an assessment of these locational tendencies from the point of view of their hosts – the regions, states, and metropolitan areas that seek to woo them. Now, we turn to ask why high tech locates where it does. Given the tendency to concentrate and then disperse, why do certain areas emerge as new centers of innovation and experience the tremendous growth that subsequent agglomeration brings? Why are others recipients of spinoffs and branch plants? Why are yet others bypassed altogether in the process? These questions are central to planners' concerns about the prospects for revitalization and job creation via the growth of high tech industry.

Regrettably, a theory of location for high technology industry does not exist. Fragments must be culled from disparate parts of location theory and other scholarship on innovation in order to begin to build a satisfactory set of explanations. The general aim of this chapter is to try to develop such a synthesis; its specific final objective is to isolate the apparently most important factors, to be tested in the statistical work that is reported in the next chapter. These fall into several categories: features of the labor force, amenities, transportation linkages, business climate, research and development spending, and defense outlays.

Traditional location theory

A well-established body of what can be called traditional location theory of manufacturing industry builds on the foundations of Weber (1929) and Hoover (1948), and has been extended in the central-place formulations of Lösch (1954). But only fragments of it are relevant for explaining the location of high technology industries.

Weber's classic formulation can be presented graphically as a location triangle, at whose corners are arrayed respectively raw materials, labor and markets. An industry locates somewhere within this triangle as determined by the relative weights of the factors. An initial location, thus determined, is subsequently affected by the growth of agglomeration economies which help to develop a locational inertia for interrelated industries. In the Löschian model, these centers of activity blanket geographical spaces, producing a hierarchy of cities in which more specialized functions and those with greater economies of scale cluster to produce primate cities such as New York or London.

These traditional models underscore the role of transportation costs in locational decision making. Empirical studies of industrial location have generally compared the costs of moving heavy inputs to the point of fabrication, with the cost of moving the resulting product to market.[1] In other words, they stress the resource and market vectors.

The problem with using these axes of the locational triangle for analyzing the siting of high tech industries is that both materials and final product are characteristically of very high value in relation to their weight in this type of industry. Transportation costs are minimal, air freight is frequently used because of its speed, and components may move long distances several times as they are incorporated into subassemblies and then into final products. These two factors, if they have significance at all, have it in the form of speed of access rather than of low cost per ton–mile; it may be crucial to guarantee the availability of components at a certain time, so that nearness to major air freight centers may be of significance. Since such industries may also involve complex and costly marketing and servicing operations by specialized and skilled personnel, easy access to major airport passenger facilities will again be essential. Our hypothesis regarding transportation factors therefore is:

(1) *High tech industry is attracted to major airports with good national and international passenger and air cargo activities.*

It is, however, the third vector, labor, that is of real significance for high tech industries. Case studies (e.g. Storper & Walker 1984, Saxenian 1985) suggest that typically high tech industrial concentrations have a sharply bipolar labor market, with a minority of very highly skilled professionals and a much larger number of quite low-paid routine process and assembly workers. (This thesis has been questioned in more recent work by Glasmeier 1986.) The former often have a critical influence on location because they are scarce, highly mobile between firms, and (because of their affluence) inclined to put a high value on quality-of-life factors. Amenities are also important to the many entrepreneurs in these sectors. Therefore, it is suggested, firms will follow them or lure them to high-amenity locations (Berry 1970, Rees & Stafford 1983, Joint Economic Committee 1982). Three different kinds of amenity need to be distinguished here: first, natural amenities such as mild winter climate, sunshine, access to oceans or mountains; secondly, availability of a wide variety of attractive housing at reasonable prices; and, thirdly, cultural amenities such as good school systems, and access to specialized cultural opportunities such as art exhibitions, concerts, opera, and ballet. These different kinds of amenities clearly have very different geographical locations: the first is found selectively in the less populous parts of the Sunbelt and in peripheral parts of the Frostbelt, the third is generally concentrated in major cities.

Since the industry also requires large numbers of lower-paid workers for labor-intensive assembly operations – and observably, in the case of Silicon Valley, has decentralized its manufacturing processes to Third World locations – it may be attracted to low-wage, weakly unionized regions. So the ideal location may not exist, especially for youthful industries which are not yet capable of separating production from research, design, and marketing activities. In terms of labor considerations, the optimal location might be a weakly unionized Sunbelt area with traditionally low wages yet with high unemployment, at the same time having an attractive climate, unpolluted air, and good cultural amenities. In reality, different portions of the production process may be drawn to sites rich in amenities in place of those well endowed with a production labor pool. The hypotheses we wish to examine regarding labor are therefore several:

(2) *High tech industries are drawn to areas with good natural amenities, in particular mild and sunny climates.*

(3) *High tech industries concentrate in areas offering attractive housing at reasonable prices.*

(4) *High tech industries are attracted to areas with educational and cultural advantages, including good educational opportunities, an array of specialized cultural services, low levels of pollution and good recreational opportunities.*

(5) *High tech industries are attracted to regions which are weakly unionized, and have low wage rates and high unemployment rates.*

Readers may wonder that we have not included the skill levels of the labor force in this discussion. Clearly, these are a prerequisite for high tech production. We do not include them in our model for two reasons. First, since we used an occupational measure to choose our high tech sectors, to use their presence as an explanatory factor in the siting of high tech jobs would involve tautological reasoning. In other words, since our high tech industries are, by definition, those with high proportions of professional and technical workers in certain occupations, of course they will be found in places with relatively high concentrations of this kind of worker. Secondly, we expect that professional/technical labor of this sort is highly mobile. There is plenty of anecdotal evidence that the vast majority of the workforce in such places as Silicon Valley have been recruited from outside the region, either directly, or via local universities, both in the past and currently; their moving costs are commonly paid for by the firm. Over time, they create a pool of available labor of this type, but it is clear that in most cases these pools did not precede the creation of new innovation centers.

Another input factor which is de-emphasized in the Weberian model but central to the earlier work by von Thünen (1826) on location is the cost of land. In the case of high tech industries, we expect that land costs are less crucial to the early location decisions of agglomerating firms in market penetration and saturation phases. Yet land availability is a critical factor for all modern industrial plants, high tech or not, because of changes in production and building technology. We did not in the end include buildable land in our model of location, largely because we could not find data to operationalize this variable. But we also concluded that this factor was more apt to operate at the intrametropolitan scale than across regions or between metropolitan areas.

The second major element of Weberian location theory, agglomeration, may also be significant for high technology industry. This industry is supposed to be highly innovative and dominated by small firms, at least at the outset. These characteristics are associated with development of specialized industrial concentrations with a high degree of reliance on external economies: the firms rely on the same skilled labor pool and scarce technical information, as well as on specialized services which are uniquely available. Silicon Valley, as described by Rogers and Larsen (1984) or Saxenian (1981, 1985) corresponds uncannily to the classic account of agglomeration economies in Marshall's *Principles of Economics* in 1890:

> When an industry has chosen a locality for itself, it is likely to stay there long: so great are the advantages which people following the same skilled trade get from near neighbourhood to one another. The mysteries of the trade become no mysteries; but are as it were in the air, and children learn many of them unconsciously. Good work is rightly appreciated, inventions and improvements in machinery, in processes and the general organization of the business have their merits promptly discussed; if one man takes up a new idea, it is taken up by others and combined with suggestions of their own; and thus it becomes the source of further new ideas. And presently subsidiary trades grow up in the neighbourhood, supplying it with implements and materials, organizing its traffic, and in many ways conducing to the economy of its material. (Marshall 1890)

Silicon Valley and similar areas, in other words, are no more than the modern equivalent of the old industrial quarter, such as the garment district of New York (Hoover & Vernon 1959) or London (Hall 1962, Martin 1966); or the small trades of Birmingham, England (Wise 1949). From this we can postulate that:

(6) *High technology industry will be attracted to areas with a high degree of internal accessibility and connectivity, as for instance areas with well-developed highway systems.*

(7) *High technology industries will be drawn to areas with a well-established infrastructure of specialized business services.*

The importance of innovation

The problem with traditional location theory – which is the basic reason why it has been found unsatisfactory – is that it is inherently static; it does not deal adequately with the essentially dynamic, unstable quality of the capitalist system as shown in the classic analyses of Marx (1867) or Schumpeter (1911, 1939, 1942). This element is of course particularly significant for relatively new industries such as most high technology industries are. We need a body of theory that accounts for the creation of new enterprises and their subsequent growth.

One way of approaching this problem is indirectly. Modern location theory, as it has developed since the mid-1970s, is essentially an account of the dynamics of the capitalist system and its spatial manifestations. The most interesting work in this tradition so far has concerned the processes of deindustrialization in older industrial regions and cities (Massey & Meegan 1982, Harrison 1982, Bluestone & Harrison 1982). This work has convincingly shown that firms in older industrial regions, faced with the challenge of competition from lower-cost regions and with a falling rate of expansion of global markets, react by a number of strategies – rationalization, capital substitution, outright closure, reorganization of productive capacity associated with closure of older plants – which all in one way or another result in losses of employment. What has been lacking so far is a parallel account of the process of job creation in newer industries.

A further step here is provided by the well-known work of Birch (1979). He suggests that everywhere and at all times, the capitalist system is marked by large numbers of new firm creations and large numbers of deaths. Typically, almost as in a biological analogy, large numbers of infant firms are born but most die soon. The death rates do not vary much from one region to another. What seem to vary are the birth rates. Some regions – Birch instances the South and West – seem to have a higher rate of new firm births than others, in particular the older industrial regions of the Northeast and Midwest.

In itself, this is simply a description; it does not provide us with explanation of the cause. One powerful body of theory that can be drawn upon here is the profit cycle theory and its antecedents discussed in Chapter 3. There, the argument was made that industries go through an evolutionary cycle in which changing business strategies lead them first to concentrate and then to disperse, as agglomerative factors are displaced by cost-cutting imperatives in the locational calculus. Evidence of such patterns was presented in Chapter 5. But the theory as developed in that model did not address the specific determinants of the crucial initial location decision, nor those dominant in disperal patterns. This issue is particularly pressing because many of the most innovative of high tech industries have not originated in older centers of research and development, as earlier theories of diffusion of innovation once posited.

To explain this tendency of innovative sectors to sprout in virgin locations, we can draw upon the Schumpeterian insights into the process of innovation (Schumpeter 1911, 1939). Innovation in this formulation, which must be clearly distinguished from invention, is the commercial application of an invention. Equivalent to entrepreneurship or enterprise, it is in Schumpeter's view the central distinguishing feature of the capitalist system. Thus the so-called industrial revolution was merely the first in a series. Based on developments in cotton textile manufacturing, in the smelting and refining of iron, and the early applications of the steam engine, it ran from approximately 1785 to 1842. A second revolution, based on railroads and the Bessemer process of steel-making, followed it from 1842 to 1897. A third, based on the modern chemical industry and the beginnings of the electrical and auto industries, ran from 1897 and was still running its course when Schumpeter wrote in the 1930s and 1940s.

Each of these innovation cycles or waves, it will be noted, was about 55–7 years long, corresponding to the long wave identified by the Soviet economist Nikolai Kondratieff in the 1920s (Kondratieff 1935). On this, in the Schumpeterian formulation, were overlaid two other cycles of shorter duration: the so-called Juglar (or classic trade cycle) of eight- to ten-year duration, and the Kitchen, occupying a mere 40 months. Much criticized on its appearance and afterwards (Kuznets 1940, 1946, 1966), Kondratieff–Schumpeter long-wave theory has enjoyed a renaissance in the recession of the late 1970s and 1980s, if only because the theory predicts just such an event (Mandel 1975, 1980, van Duijn 1983). Most students of the subject now seem to accept the existence of a long wave of some 55–60 years' duration, though they argue about causes. Some, notably the German economist Gerhard Mensch (1979), argue that, throughout capitalist history, innovations have significantly bunched at certain points in time, at about 1764, 1825, 1881, and 1935 – just when the theory would demand. Others doubt this, arguing that extraneous and contingent forces are responsible (Mandel 1975, 1980, Clark *et al.* 1981, Freeman *et al.* 1982).

This whole body of theory, again, is almost completely aspatial. But it is not difficult to think of elements of a spatial theory of innovation. There are suggestions in the economic history literature that innovative industry tends to develop in areas different from the previous industrial concentrations, because the latter suffer from some hardening of the innovative arteries. An example is the classic early work on the English medieval and early modern woolen industry, which shifted first to Suffolk, then to Norfolk, then to the West Riding of Yorkshire (Clapham 1910, Carus-Wilson 1941) as the former leading region failed to innovate. Similarly, Checkland (1975) has suggested that Glasgow's economic decline occurred because of its concentration on custom shipbuilding and its failure to develop incipient new industries such as vehicles or aircraft. For the USA, both Chinitz (1960) and Markusen (1985a) argue that oligopolistic domination of local resource markets explains the relatively poor growth performance of cities such as Pittsburgh (steel) and Detroit (automobiles). In contrast, G. C. Allen's early work on Birmingham, England (Allen 1929) suggests that the city's success lay in the continuing innovative capacity of its small workshops. More recently, Lloyd and his coworkers have suggested that Manchester's diversified industrial economy produced a higher rate of innovation than Merseyside's (Dicken & Lloyd 1978, Lloyd & Mason 1978).

The nature of such an innovative environment is not, however, easy to pinpoint. Except that it is unlikely to be present in an old and established industrial region, the distinguishing features are not very clear. It has been suggested that a favorable business climate, represented by little regulation and low rates of taxation, will allow enterprising individuals to innovate freely and that this is one of the main sources of the rise of the American Sunbelt (Perry & Watkins 1978, Birch 1979). This might be tested empirically by using proxies for business climate. Thus our hypothesis:

(8) *High technology industries are drawn to places which have an antiregulatory, free-enterprise ideology.*

The geography of research and development

The business-climate hypothesis, however, ignores or sidesteps the fact that much innovation, especially in the high technology industry that concerns us here, arises from organized research and development (R&D). Though a great deal of this may well take place in the laboratories of quite small enterprises, some may be organized through multi-plant, multinational enterprises which presumably have a great deal of freedom to choose their location. The difference lies in the fact that as in other activities, so especially in R&D, the small firm may be extremely sensitive to economies of agglomeration. And these may not lie merely in the presence of other similar firms, but also in the presence of major centers of research (universities, government research establishments) which both make the basic scientific discoveries that can be applied commercially, and also generate demands for specialized products or processes.

A significant literature now exists on the geography of R&D both from the United States and from Great Britain. Malecki's work (1980a,b,c, 1981a,b), which is the most considerable empirical analysis, makes a fundamental distinction between R&D generated by private industry and that generated by government. Though no less than 70% of federally funded R&D was done by private industry, its geographical distribution was quite different from R&D done by private industry on its own account.

Malecki found that R&D as a whole was quite strongly concentrated in a quarter (58 out of over 200) of Standard Metropolitan Statistical Areas. He classified these into three quite different types. The first was associated with the headquarters of private industrial firms. These places had a high percentage of manufacturing employment but very low levels of federal or university research, and most (with a few exceptions like Minneapolis–St. Paul, Wichita or Seattle) were located in the traditional American manufacturing belt. The second had high concentrations of R&D associated with non-headquarters manufacturing, and again – with a few exceptions (Atlanta, Kansas City, Dallas–Fort Worth) – was mainly found in the manufacturing belt. The third consisted of the true innovation centers, mainly university cities and centers of federal research, many of which were located outside the manufacturing belt. The few exceptions tend to have diversified economies and some high technology industry. Malecki distinguishes two subtypes: one university-based (Austin, Texas; Lincoln, Nebraska), the other based on federal R&D (Washington, DC; Huntsville, Alabama). The real distinction, it can be suggested, is between these two subtypes and the other two categories, which largely carry out non-innovative research in the interests of maintaining the position of well-established industry (Malecki 1980a).

In his later work, Malecki makes an important discovery that federal research support is completely dominated by defense. No less than 77.7% of all federal R&D support to industry goes to two industrial sectors, aerospace and electronics, and 58% comes from the Department of Defense. Further, this kind of R&D is highly concentrated geographically: 61% of all R&D laboratories in aerospace are in the Los Angeles area, while five metropolitan areas (Boston, New York, Philadelphia, San Francisco, Los Angeles) account for much of the electronics research (Malecki 1981b).

It is because of this that there is a significant difference between the distribution of industrial R&D and that of federal R&D. The latter is highly concentrated into cities without much industrial research

(Albuquerque, New Mexico; Knoxville, Tennessee; Pensacola, Florida; Seattle, Washington), while little of it goes into cities with industrial research concentrations (Buffalo, Cleveland, Indianapolis, Kansas City, Milwaukee). Malecki identified only 41 SMSAs with location quotients for either kind of R&D of 1.0 or more; of these, only 11 scored on both lists. The major centers of R&D – Boston, Washington DC, Houston, Los Angeles, San Francisco – appear here together with a few smaller, more specialized places like Dayton, Ohio; Lafayette, Indiana; Madison, Wisconsin; Huntsville, Alabama; and Santa Barbara, California (Malecki 1980a).

Malecki's work indicates that R&D is strongly concentrated in a few places. What is difficult, however, is to distinguish continuing process research from the innovative R&D that contributes to new industrial development. The suspicion is that it is mainly the federal R&D that has at least the potential quality to achieve this, perhaps through commercial spinoffs from military R&D. The well-known experience of Silicon Valley, which in its early years was basically an industrial offshoot of electronics research at Stanford University with heavy Department of Defense backing, illustrates this perfectly (Saxenian 1985). So does the more recent experience of the infant biotechnology industry in the San Francisco Bay Area, which appears to be an offshoot of research at the University of California, San Francisco medical campus (Feldman 1985).

Limited evidence from Great Britain also suggests a link between government R&D concentrations and the locations of new high tech industry. The analyses of Buswell and Lewis (1970) and of Howells (1984) both suggest a heavy concentration of such research in Southeast England; while the work of Oakey *et al.* (1980) indicates that this region easily leads in number of industrial innovations. Older industrial regions (such as Wales, Scotland, the Northwest and West Midlands) scored low both on R&D and on industrial innovation. The marked concentration of high technology industry along the so-called M4 Corridor – actually a series of locations along the South Wales freeway west of London – appears to be related to the initial presence of government research establishments in the area, as well as to the attractions of nearness to Heathrow for American firms and the generally high environmental quality (Breheny *et al.* 1985).

Three alternative hypotheses emerge from this discussion: one concerned with industrial R&D, a second with fundamental research, the third with defense:

(9) *High tech industry will be drawn to centers of industrial R&D which will tend to locate close to the headquarters of major industrial corporations.*

(10) *High tech industry will be generated in locations with concentrations of federally funded fundamental scientific research.*

(11) *High tech industry will be generated in areas with high concentrations of defense spending.*

Some conceptual issues

These 11 hypotheses formed the basis of our empirical work both on the distribution of high technology industry and on the factors in its generation and growth. Before going on to report our tests of these theoretical notions, we must reflect on a number of conceptual issues confronting this exercise.

First of all, we are dealing here with an historical, dynamic phenomenon. As our previous chapters showed, high tech industries have very different business strategies at different historical moments. Collapsing the locational priorities of firms and sectors in very different stages of evolution into an ahistorical, cross-sectional model contradicts much of our conceptual work. In our aggregate model, we are testing whether factors such as low wages and a good climate are important in explaining high tech job shifts. But in reality, climate is pulling some, probably the more "youthful" sectors, its way, while low wages are pulling others, the market saturated and rationalizing sectors, another. Drawing conclusions, then, from the aggregate patterns does not adequately tell us about the longer-run tendencies of individual high tech sectors as their markets evolve.

Secondly, the factors just cited are not equally differentiated at every geographical scale. As the brief note on buildable land suggests, many locational factors are more powerful determinants of choices at one spatial scale than another. For instance, wage rates are less apt to vary across metropolitan areas or within states than they are between states and regions. Similarly, spending on education and voting patterns will vary more strikingly within metropolitan areas than among them. We were able to test the locational hypotheses only at the intermetropolitan scale. Readers should not assume, therefore, that factors found insignificant at one geographical scale are also insignificant at others.

Thirdly, unrepeatable historical anomalies or crises may have in fact been the initiators of a string of subsequent location decisions. It may be that the full history of high tech location begins with historical factors such as the War Production Board's need to build up West Coast engineering capacity during World War II, or with Lyndon Johnson's preference for siting the space program in his native South. A set of hypotheses such as those we have just posited cannot easily incorporate these momentous decisions nor follow through the chain of events set off by them.

Finally, locational orientation responds to secular changes in the economic environment. The construction of the interstate freeway system, the advances in telecommunications technology, the changing size, function and atomization of the family; these and many other structural changes have had a dramatic impact on industrial location and human settlement patterns in general. A comparative look at the endowments of metropolitan areas which are presently gaining (or losing) high tech plants and jobs does not adequately reveal the role of such society-wide innovations in the unfolding of recent high tech locational patterns; that would require a longer, time-series analysis of shifts over time, controlling for all such changes.

Yet given all of these caveats, it is still worthwhile using the evidence from our very fine-grained high tech location data to detect the relative strength and weakness of these hypothesized factors in recent high tech locational patterns. It is to this task that we now turn.

Note

1 See, for instance, the debate between Heckman (1978) and Karlson (1983) on the location of steel.

9
Where and why high tech locates

Applying theory to data

IN THIS FINAL part of our analysis, we try to bring the *where?* and the *why?* questions together. We use the hypotheses, derived from the theoretical discussion in Chapter 8, to identify factors that might explain the high tech geographical distributions that were described in Chapter 7. We embody these factors in a multiple-regression model to see how far, collectively and partially, they might explain those distributions. We do this twice, to try to explain first the 1977 patterns and, secondly, the 1972–7 changes. These, we argue, represent two different sets of processes: the 1977 distribution represents the cumulative long-term effect of decisions over the whole of a product–profit cycle and even over more than one, while the 1972–7 changes represent the effect of forces working over a shorter (and, it must be admitted at the outset, possibly atypical) recent timespan.

The variables fall into four reasonably well-defined groups. One consists of variables describing the characteristics of the local labor force: wage rate, unionization rate, unemployment rate and percent black. A second contains variables that measure metropolitan amenities: climate, housing prices, and educational options. A third set of variables consists of access features: freeway density, and access to a major airport. A fourth group is intended to capture the agglomerative features thought to be important for high tech industries: the presence of major business headquarters, a wide range of business services, and the amount of fundamental research funding that the area receives. Finally, we include a special variable which also measures R&D intensity: defense spending.

The analysis is carried out at a spatially disaggregated scale, for the 264 SMSAs included in the SMSA-level analysis of Chapter 7. And, having reported results in aggregate, we go on to make separate analyses for different SMSA size classes and for different groups of industries. In the latter, our aim is to return to the insights provided by product–profit cycle theory, asking: are the explanatory variables different in their impact for youthful, middle-aged and mature industries?

Our results are summarized at the end of the chapter. We should not anticipate them here. But we can encapsulate them in a sentence: the overall level of explanation is good; it is, however, better for 1977 distributions than for 1972–7 change; and, as to the impact of individual factors, it yields surprising results that contradict conventional wisdom.

A model of high tech job and plant location

The model we constructed to test our hypotheses about the location of high tech industries was designed to operate at the metropolitan level. We chose to use the smallest unit of aggregation possible given data availability. State level distributions, for reasons made clear in Chapter 7, are too highly aggregated to detect dramatic variations within state boundaries. On the other hand, it proved impossible to find for the county level sufficient data on area characteristics. Metropolitan areas more or less correspond to labor markets and have the added feature that they transcend state boundaries where appropriate. The choice of the metropolitan level of aggregation meant that a number of hypotheses which we would have liked to test – for instance, the voting complexion of the population (a crude measure of business climate) and variations in local taxation and spending patterns – could not be pursued, since the variation occurs across the metropolitan area rather than between metropolitan areas.[1]

Exogenous variables

We incorporated 13 variables in our model, representing the metropolitan features we hypothesized to be associated with high tech job and plant location. Three of these are traditional features of a local labor market: wage rates, unionization rates and the unemployment rate. The wage rate, the average manufacturing wage for the metropolitan area, which varied from approximately $4 to $8 per hour in 1977, was hypothesized to be negatively related to high tech job and plant locations, especially as patterns of dispersal were taking place over the period. The unionization rate, the percent of the labor force unionized, was also hypothesized to be negatively related to high tech job and plant locational patterns. The unemployment rate, as a measure of surplus capacity in area labor markets, was expected to be positively related to high tech location, again particularly in the redistribution of activity in the mid-1970s.

We also included a fourth, less conventional labor variable: the presence of minorities in the metropolitan population, approximated here by percent black.[2] This variable, we hypothesized, was negatively related to location, as a proxy for racist attitudes on the part of employers and fellow employees. Recent reports exploring Census data have suggested in general that manufacturing relocation has spurned places with high proportions of blacks in the workforce for similar places which are predominantly white; this has been particularly true in the rural South.

We chose not to include a variable which would have captured differences in skill levels and occupational composition of the area-wide workforce. While we believe that there is a strong correlation between the proportion of the area-wide workforce that is professional/technical in nature and the incidence of high tech activity, we had already used this variable to identify our high tech sectors to begin with. To test the relationship in this manner would be an exercise in tautology. Furthermore, our reading of the literature on high tech location and our own previous case studies suggest that high tech labor pools have been created anew in locations like Silicon Valley, rather than drawing high tech activity to a previously existing labor force. More relevant, then, are the amenities which are hypothesized to attract high tech personnel to an area.

A second set of variables thus represents metropolitan amenities. These are envisioned to operate both as factors attracting a relatively mobile, professional and technical workforce, and as inducements to entrepreneurs. The climate variable, an index of climatological characteristics which generally favor mild over volatile climates, was hypothesized to contribute positively to high tech presence. A housing price variable, the average sales price in the metropolitan area, was hypothesized to be negatively related to high tech location, especially to shifts in the 1970s. Educational options, a measure of post-secondary educational programs, is included because this type of amenity is generally thought to be important to a professional/technical workforce, for both personal and family reasons.[3]

The third set of variables can be termed "access features." Freeway density – the number of freeway miles in the metropolitan area divided by its land area – was hypothesized to be positively related to high tech activity. Airport access, an index which rates area airports from 0 to 4 on the basis of the amount of passenger activity they account for, was hypothesized to be positively related to high tech activity, not only for air cargo shipment, but for the type of business travel important to an innovative industry. Because high tech industries generally produce relatively high value-added to weight products, the accessibility to truck and air transport was hypothesized to be more important than access to water or rail transport facilities. No variable was therefore included for the latter type of transportation access.[4]

Three variables were included to capture the agglomerative features hypothesized to be important to innovative high tech industries. The number of Fortune 500 firms with headquarters in the area was hypothesized to be positively related to high tech activity, since headquarters generally host both research and development activities and centers of decision making. A second agglomerative variable was business services, the percentage of the area employment accounted for by those activities such as accounting, consulting, advertising, research and development laboratories, data processing and computer software services. These were hypothesized to be positively related to high tech development, representing the external economies available to small, new firms locally. A third variable, university-based receipts of research and development funding, both from private and public sources, was hypothesized to be positively related to the presence of high tech activity, since it approximates the presence of major research universities which are commonly believed to have been so critical in the evolution of places like Silicon Valley and Route 128.[5]

We also included a less conventional measurement of research and development activity: defense spending per capita, a measure of the infusion of federal prime contracts awards. This was hypothesized to be positively related to high tech activity, since – as we have argued above – the role of defense spending in subsidizing research and guaranteeing a market for high tech products had been substantial; the inclusion of this variable was designed to test whether this role extended to the spatial realm as well.

The endogenous variables

We were interested in testing the contribution of these metropolitan characteristics to two different aspects of high tech activity: the overall locational pattern as observed in 1977, and the changes in that pattern between 1972 and 1977. The former would be a test of the extent to which contemporary place features are associated with the historically constructed high tech complex across regions. The latter would capture the degree to which these features had been associated in the most recent period with high tech growth and decline.

To test the degree to which these variables "explain" the 1977 pattern of high tech distribution across metropolitan areas, we formulated two versions of the endogenous variable. One was the number of high tech plants located in that metropolitan area according to the 1977 Census of Manufactures. The other was the level of metropolitan employment in high tech industries in 1977, as estimated by the procedure outlined in Chapter 5. To test the degree to which these variables were associated with shifts in plants and employment in the mid-1970s period, we formulated two additional endogenous variables: the absolute change in the number of high tech plants in the period 1972–7 and the net change in employment during the same period, estimated by the same procedure.

It is important to keep in mind that we are building here a model which is designed only to test cross-sectional variations, at one moment in time, in high tech locational patterns. In reality, the current pattern of high tech location is the product of an evolutionary process which may have had very different determinants in different periods of time. Indeed, our own theoretical work (reported in Chapter 6) suggests that agglomerative factors are more important in anchoring the innovative portions of high tech activity in the early stages of an industry's development, while cost factors (land, labor, housing, etc.) may become relatively more important in later periods. In addition, both factors may be operating simultaneously. As the more routine activities are spun off from the innovative firm's home operations, these may be drawn to locations with entirely different features than those that dominated, and continue to dominate, the pull on the innovative center of the same firm's operations.

Thus the tests we are performing on this model are relatively primitive compared to the power of the theory offered above. Were we, for instance, able to discriminate among plants in any single SIC on the basis of experimental versus routine manufacturing activities, we might find that some variables explain well the siting of the former while others are dominant in the latter. Furthermore, if we could construct historical series on both the exogenous and endogenous variables in our set, we would be able to test the hypotheses that different variables are sequentially more or less important in the locational orientation of high tech industries during the unfolding of their developmental process.

Thus the model incorporates rather snapshot-like relationships which must be carefully interpreted. The reader must continually keep in mind that we are using a cross-sectional model to test what are really developmental relationships, a common practice but one which cannot therefore be used as a forecasting device. For instance, if we detect a significant, positive contribution of per capita defense spending to the 1977 variation in high tech location, this does not necessarily tell us that this pattern will hold true in the future nor that it is driving present additions to high tech plant and employment. The model is designed modestly to test only whether at a given time, 1977, the differentials across all metropolitan areas in labor, amenities, transportation access, agglomeration features, and so on, are associated with differentials in high tech job and plant incidence.

The formulation of the model with the second set of endogenous variables, the absolute levels of net change in plant and job incidence, is an attempt to go some way beyond the simple cross-sectional model by asking whether patterns observed in the aggregate also held true for the net additions to plant and employment in the mid-1970s period. A finding that certain variables were more or less significantly related to job and plant *change*, as opposed to incidence *per se*, can be taken as an approximation of the relative power of factors operating in the contemporary period rather than in the longer run. It can then be inferred that high tech industries, which as a group are presumably somewhat more mature in the mid-1970s than they were over the longer postwar period, are at this time relatively more responsive to certain metropolitan features than to others which may have predominated in earlier stages. But it is important to keep in mind that this inference is still drawn from a cross-sectional test, not a dynamic model. We have not, for instance, tested whether changes in high tech location in the 1970s were associated with *changes* in metropolitan features in the same period, only whether they appear drawn to places with differentially high or low labor costs, amenities, and so on.

We included the same set of variables in each of the two groups of equations. However, we did hypothesize that the amenities and agglomeration variables would be relatively more important in explaining the overall distribution of activity than they might be in the shifts of the mid-1970s, a period when dispersion had clearly begun from agglomerative centers. On the other hand, we expected that factors which had more closely to do with cost of production (the labor and transportation access variables) and the defense-spending variable, which is a revenue-side feature and presumably a proxy for "market," would be relatively more strongly associated with mid-1970s shifts than with the overall historically evolved distributional patterns.

Model formulation

These hypotheses were incorporated in four versions of the basic model of high tech location.

$$P_m = f(w_m, u_m, c_m, h_m, O_m, f_m, A_m, F_m, s_m, r_m, d_m, b_m) \tag{1}$$

$$E_m = g(w_m, u_m, c_m, h_m, O_m, f_m, A_m, F_m, s_m, r_m, d_m, b_m) \tag{2}$$

$$\delta P_m = h(w_m, u_m, c_m, h_m, O_m, f_m, A_m, F_m, s_m, r_m, d_m, b_m) \tag{3}$$

$$\delta E_m = j(w_m, u_m, c_m, h_m, O_m, f_m, A_m, F_m, s_m, r_m, d_m, b_m) \tag{4}$$

where:

P_m = number of plants in the metropolitan area, m
E_m = number of jobs in the metropolitan area, m
δP_m = net change in number of plants, in metropolitan area, m
δE_m = net change in number of jobs, in metropolitan area, m
w_m = average metropolitan manufacturing wage rate
u_m = rate of unionization in the non-agricultural and non-military workforce
c_m = index of climatological conditions in the metropolitan area

h_m = average sales price of a home in the metropolitan area

O_m = metropolitan rating on an index of available educational options at the post-secondary level

f_m = number of freeway miles in the metropolitan area divided by the land area

A_m = metropolitan rating on an index of airport accessibility

F_m = number of Fortune 500 headquarters located within the metropolitan area

s_m = proportion of the local labor force engaged in business service industries

r_m = level of university private and publicly sponsored research and development funding within the metropolitan area

d_m = per capita defense spending on prime contracts with a value of over $10,000

b_m = percent of the metropolitan population which is black

In each version, variables which have been scaled – by population (minorities, defense spending, R&D), workforce (unionization, unemployment) or area (freeway density) – and variables which are area-wide averages (wage rates) are expressed in lower case while those which are not scaled (Fortune 500, airport access, educational options) are expressed in capitals. This is an important feature of the model, since we did not want sheer size of metropolitan area to influence the outcome, especially since our endogenous variables were not scaled. Only in those cases where the size of the city would not presumably cut down on the access of high tech firms to the features modelled (e.g. climate, Fortune 500) was it appropriate not to scale by an indicator of metropolitan size.

Model specification and estimation

The data What we have, then, is a relatively large-scale, cross-sectional model of high tech location across metropolitan areas. Specifying and estimating a model of this size requires an extensive data set. We were able to locate data on all of our variables, although the quality of the data must be carefully considered. A major factor in both the formulation and specification of the model was the availability of data on the metropolitan variables. Data on the number of plants was available to us from the Census of Manufacturing tapes. We have already discussed, in Chapter 6, the procedure used to estimate employment levels for counties; these were aggregated to get metropolitan totals for 264 metropolitan areas in the USA. This is a significantly larger set of metropolitan areas than is normally used in metropolitan-level empirical work. Indeed, it is essentially the entire set of US metropolitan areas, with the exception of those which have only been designated since the 1980 Census and two for which we could not get sufficient data on our set of characteristics.

We chose to include the entire set because we did not want to eliminate smaller metropolitan areas which might be pivotal in the redistribution of high tech activity in the period studied. Overall, this was a set of metropolitan areas whose growth rates exceeded those of the nation as a whole during this period, so that their elimination might seriously bias the results. The cost of including all metropolitan areas was that it constrained us in choice of variables to some extent, since fewer data sets on metropolitan characteristics cover the entire US group.

A description of the data sets used for each exogenous variable and the sources, as well as additions made to the sets ourselves, is included in Appendix 3. The reader should note four caveats. First, most but not all of the variables are continuous series. We tried to employ these as far as possible, expecting that they would improve the estimation results. The exception is the airport access variable, with rankings of 0 to 4. It should also be noted that the educational options and climate variables are indices, so that their accuracy is dependent upon the procedures used in aggregating different features of places into a single index.

Secondly, it proved impossible to gather all series for the same year. It would have been optimal to use data for 1977 in all cases, since that was the year for which we were testing the contribution of these variables to variation in high tech activity. However, it was not always possible to get data for all of our metropolitan areas for 1977. We chose that year in all cases in which we could get a full data set on all of our metropolitan areas. In most cases, these data sets existed for 1977 or 1978. The exceptions were the black proportion of the population (1970), housing prices (1976) and unionization rates (1976). In these cases, we chose to assume that the relative differentials across metropolitan areas had not changed substantially in the intervening year or years. This may not be reasonable in the case of percent black, yet we could think of no bias in black patterns of migration in the early 1970s which would seriously prejudice our results in using 1970 proportions as a proxy for 1977 proportions.

Thirdly, one variable, unionization rates, was only available at the state level. In most states, these rates may be more or less the same across metropolitan levels due to right-to-work laws at the state level. However, actual rates may vary within states, especially larger ones. For instance, it is likely that unionization rates are higher in the southeasterly Texas cities like Beaumont and Houston which have older traditions of oil, steel, shipbuilding and shipping-related activities than in Dallas, Austin, or western Texas. We chose to keep this variable because it captured an important business climate feature, yet the possibility of bias from this source must be kept in mind.

Finally, and perhaps most importantly, bias could be introduced into the estimation procedure because of the fact that some metropolitan areas are contiguous. From the point of view of data issues, this is a problem only with respect to those variables which represent industries and/or services which might fall on one side or another of a metropolitan boundary but be accessible to both. Airports are perhaps the most outstanding example. Silicon Valley, for instance, relied not only upon the San Jose airport, which is incorporated in its metro airport access index, but also upon San Francisco and possibly Oakland airport as well, which are not. Anaheim SMSA may be in the same position regarding the Los Angeles International Airport. Other variables in which this may be a problem include educational options, Fortune 500 headquarters, business services and university research and development spending. We could not figure out a way of correcting for this possible inclusion of bias. We return to this point in the discussion of spatial autocorrelation below.

Despite these caveats, which the reader should keep in mind, we were able to assemble a large data base on the entire set of US metropolitan areas for each of our exogenous variables. The data base thus provides very detailed information about characteristics of the labor force, transportation access, amenities, agglomeration factors, defense spending and racial composition. With this, we were able to proceed with a test of our hypotheses about location.

Model specification Inevitably, there were a number of technical problems involved in developing a proper functional form for our model. These are described in Appendix 4. Thus several of the variables were specified in a logarithmic form in order to reduce the enormous differences in scale between the many small metropolitan areas and the few very big ones. Some of the original variables, discussed in the notes at the end of this chapter, were removed because they proved to be highly correlated with others. But we could not remove all the multicollinearity in the model, and this may bias our results in a way that cannot be determined. A reverse problem is that, almost certainly, some relevant variables are missing because we could find no good data for them: buildable land and the local tax burden are the most important that we are aware of, but we may have missed others. Finally, there could be a problem of spatial autocorrelation: results for two adjacent metropolitan areas may not be truly independent of one another, but the literature gives no guide as to how this problem can be overcome. All these limitations may have affected our results; the reader should be aware of them.

Testing the model for high tech location

The first set of tests involved estimating the model for the year 1977, for both plant distribution and job incidence. The results are shown in columns 3 and 4 of Table 9.1. The parameter estimates are shown, along with their signs, t-statistics and a symbol which indicates whether the coefficient is significant at the 1% (**) or 5% (*) levels. Significance in this case means that we can reject the null hypothesis that the sign is not the expected one.

The labor force features performed poorly in our tests. Both wage and unionization rates had perverse signs, indicating that high levels of each were positively associated with high tech and job plant locations as they existed in 1977. However, neither coefficient was significant at the 95% level of confidence. The unemployment rate did exhibit the hypothesized sign for plants, although not for jobs, but even in the former case the null hypothesis could not be rejected.

Poor findings on this set of labor force variables could be due to the fact that the causal relationship is operating in the opposite direction − that high tech activity in itself drives up wages, absorbs surplus labor, and encourages or at least tolerates unionization. Evidence from other studies suggests that the wage rate effect, due especially to the professional/technical portions of the workforce, may indeed operate in this manner, but anecdotal evidence also suggests that high tech firms are very adamantly opposed to unionization. Thus, it could be that agglomeration factors may override traditional labor cost and quality variables, at least during initial stages of development only. If so, then we

Table 9.1 Factors associated with metropolitan high tech job and plant location.

Locational variable and category	Hypothesized sign	Estimated dependent variable coefficients			
		Plants 1977	Jobs 1977	Plant change 1972–7	Job change 1972–7
Labor					
wage rate	−	0.06 (1.23)	0.10 (1.28)	−0.08 (−1.05)	−0.08 (−1.44)
unionization rate	−	0.02 (3.20)	0.02 (2.57)	0.01 (0.48)	−0.01 (−0.88)
unemployment rate	+	0.02 (0.79)	−0.06 (−1.49)	−0.06 (−1.38)	0.00 (0.05)
Amenities					
climate index	+	$0.12e^{2}$** (2.46)	$0.25e^{2}$* (3.08)	$0.15e^{2}$* (1.74)	$0.14e^{2}$* (2.24)
housing price	−	$-0.65e^{5}$ (−0.99)	$-0.23e^{4}$** (−2.19)	$-0.94e^{5}$ (−0.87)	$-0.14e^{5}$ (−0.17)
educational options	+	0.02** (6.15)	0.02** (4.50)	0.02** (3.47)	0.01** (4.19)
Access features					
freeway density†	+	0.24** (4.56)	0.27** (3.30)	$0.46e^{2}$ (0.05)	−0.02 (−0.30)
airport access	+	0.16** (3.25)	0.08 (0.94)	0.01 (0.15)	0.11* (1.80)
Agglomeration					
Fortune 500†	+	0.21** (2.41)	0.13 (0.96)	−0.22 (−1.55)	0.18* (1.75)
business services†		+0.67** (5.86)	0.72** (3.98)	0.50** (2.67)	0.51** (3.75)
university R&D†	+	−0.06 (−2.00)	−0.08 (−1.74)	−0.04 (−0.92)	0.01 (0.37)
Socio-political					
defense spending†	+	0.05 (1.38)	0.12* (2.23)	0.09* (1.65)	0.10** (2.42)
percentage black	−	−1.16* (−1.92)	−0.79 (−0.82)	−2.24* (−2.25)	−2.09** (−2.90)
Constant		1.27* (2.23)	6.01** (6.62)	8.17** (8.70)	2.15** (3.16)
R^2		0.71	0.52	0.23	0.45
F-statistic		47.0	20.7	5.7	15.7
number of cases		264	264	264	264

Figures in parentheses are t-statistics.
† Variable logged.
* Indicates rejection of the null hypothesis that the sign is not the expected one, at the 95% level of confidence (one-tailed test).
** Indicates rejection at the 99% level.

would expect the next set of regressions on recent change to provide better results on these variables.

The amenities variables, on the other hand, appear to be more satisfactory in explaining high tech job and plant location. The estimated parameters for both climate and educational options showed the hypothesized sign and were significant at the 1% level. Housing prices also exhibited the expected sign, but were significant only in the case of job location. We can quite confidently confirm that good amenities have been associated with the historical evolution of high tech complexes.

Access features were also found to be significant contributors to explained variation in high tech job and plant locations in 1977. Both freeway density and airport access were positively associated with high tech plant siting, and the former was also significant in the jobs distribution, all at the 1% level. However, the coefficient for airport access in the jobs regression could not be confirmed as significant, suggesting that airport access may be less important for the siting of larger, employment-intensive plants than for smaller, experimental, or innovative ones. This would suggest that the business travel aspect of airports may be a stronger attractional force than the air cargo services they offer.

Agglomeration features also turned out to be rather important, at least in explaining variations in plant location. The presence of Fortune 500 headquarters was a significant and positive contributor to explained variation in plants, although not jobs. This again suggests that smaller, innovative and/or experimental operations are more apt to cluster around headquarters complexes than the more routinized, mass production facilities. Business services, on the other hand, were found to be positively and significantly (at the 1% level) associated with high tech location for both jobs and plants. The university research and development variable produced the opposite sign from that expected; the main reason seems to be that some fairly small and isolated centers, with little high tech activity, attract large amounts of R&D per head of population.

The defense spending variable was unique in the test of high tech location in 1977 for being a significant explanatory factor in job but not in plant distribution. In both cases, the sign of the parameter generated was positive, but could not be confirmed as significant in the case of plant location. This could be a sign that defense spending plays a more powerful role as a procurement factor in high tech industries than as a locator of research and development activity. Finally, the minority share of a metropolitan area's population did turn out to be negatively related to high tech job and plant distribution in 1977, although the null hypothesis that minority composition does not matter could be rejected only in the case of plant location.

Overall, then, we found that seven variables contributed as expected to variations in plant sitings in 1977 and six to variations in job distribution. Housing prices and per capita defense spending were significant in the jobs case but not for plants. On the other hand, airport access, the presence of Fortune 500 headquarters and minority presence figured significantly in plant location but not for jobs. In general, the labor force variables were disappointing, as was the university research and development variable; these were the only cases in which parameters generated displayed the opposite sign from that hypothesized.

The overall level of explained variation was also quite impressive in this set of regressions, especially for such a large cross-sectional effort. The set of exogenous variables explained 71% of the variation in plant distributions, and the same set explained 52% of variation in job distributions. It appears that the factors we have identified did account for a major portion of the variation in high tech plant and job distribution across metropolitan areas in 1977. Since the locational pattern as it existed in 1977 was the product of a longer-term historical evolution, predominantly though not entirely postwar in duration, we might conclude broadly that amenities, access and agglomeration factors were important features in the high tech locational process; more so, certainly, than conventional labor force attributes.

We also took a look at the outliers to see what types of metropolitan areas defied the model's expectations. For high tech job distributions in 1977, the worst outliers were a group of peripheral locations whose high tech employment in 1977 was over-estimated by the model. These include Columbia, Missouri; Pueblo, Colorado; Anchorage, Alaska; Lawton, Oklahoma; Killeen, Texas; Great Falls, Montana; Lafayette, Louisiana; and Honolulu, Hawaii – all small to medium-sized SMSAs outside the traditional manufacturing belt. Of the top ten outliers, only two were underestimated by the model: Lakeland, Florida, and Johnson City–Bristol–Kingsport, Tennessee/Virginia. The predominance of over-predicted SMSAs in this group shows that a place may possess the requisite qualities and still not attract high tech jobs.

Testing the model for high tech change

A set of equations was tested to determine the degree to which the net changes in high tech plants and jobs in the period 1972–7 could be explained by the same differentials in metropolitan characteristics. The parameters estimated and their t-statistics are presented in columns 5 and 6 of Table 9.1. Overall, the level of explained variation fell in this set of regressions. We were able to account for 23% of the net change in plants in the mid-1970s and 45% of the net change in employment. Nevertheless, the signs and significance of individual parameters demonstrated some striking differences from the previous set of regressions, suggesting that certain factors began to play a more prominent role in high tech orientation in the more recent period, while others receded in importance.

The labor force variables were again not significant. However, this time the wage rate parameter did exhibit the expected sign for both plant and job change, and the unionization rate showed the expected negative sign with respect to job change. This could mean that during the 1970s conventional labor force features were becoming more important in affecting new high tech location decisions and in selectively shaping plant closings than had been true previously; but this is highly speculative, since the effect is only marginal and is not significant. The null hypothesis could not be rejected in the case of employment either, although in this set of regressions the rate had the expected sign for job change but not plant change: the flip side of what occurred in the 1977 regressions.

The amenities variables were again relatively more important, suggesting that the features of an area that are attractive to a professional/technical workforce and managers continued in the 1970s to be associated with net changes in high tech activity. However, the role of amenities was somewhat less notable than it had been in explaining the overall distribution of activity. The climate variable was significant in both cases, but could only be confirmed at the 5% level instead of at 1% as previously. The housing price variable again had the expected sign but was no longer significant even at the 5% level. Educational options was the only one of the amenities variables which remained positively and significantly related to variation in high tech growth at the 1% level of significance.

Access features appeared to lose their power in attracting high tech development in the mid-1970s. Freeway density was not significant in explaining changes in either plant or job distributions and, in the case of jobs, the parameter generated was actually negative. Airport access was no longer significant in explaining plant siting, although the sign of the parameter remained positive. But airport access did become a significant explanatory factor in job change. Thus the air cargo function of airports may have been growing in importance as a factor in mid-1970s high tech location, while the business travel function was declining.[6]

Agglomeration factors in shaping mid-1970s growth were also somewhat less straightforward than in explaining the overall distribution. The parameter generated for university research and development continued to display the opposite sign from that hypothesized. Fortune 500 headquarters were not only not significant in explaining plant change during the period, but the sign of the parameter generated in the regression was negative. This would suggest that in the most recent period, the dispersal tendencies in high tech industry had begun to overtake the agglomerative tendencies. However, Fortune 500 headquarters turned out to be positive and significant (at the 5% level) contributors to job change in the period. A generalization about the diminishing role of headquarters attraction is thus impossible to draw.

Business services, on the other hand, were as significant and positively related to both plant and job change as they were in explaining overall distribution. This suggests that the presence of a complex of business services is strongly and continuously associated with high tech activity. However, it is possible that this relationship is also one in which the direction of causality is reversed. High tech activity may be drawing business services around it rather than responding to pre-existing pools of business services.

The socio-political variables, defense spending and percent black, showed the most distinctive tendency to become relatively more important in explaining mid-1970s changes than they were even in the overall distribution. Defense spending per capita was positive and significant for both plant (at the 5% level) and job change (at the 1% level). The contrast between its role in this set of regressions and those for overall plant and job distribution suggests that as a factor shaping high tech location, defense spending became more powerful in the period. This is somewhat remarkable since the five years scrutinized were ones of relative military build-down. That is, despite lower relative levels of defense spending than had prevailed in the previous Vietnam era, metropolitan variations in those levels became relatively more prominently associated with net changes in high tech activity.

The presence of minority populations as a factor associated with low levels of relative high tech plant and job change is also striking and disturbing. The parameters generated were negative and significant for both plants (at the 5% level) and jobs (at the 1% level). Both the size of the parameters and the strength of the statistical relationship show that net growth in high tech activity shunned metropolitan areas with high proportions of blacks. This is undoubtedly a product of differentially concentrated plant closings, heavy in northeastern cities with large black populations, as well as plant openings or additions skewed toward Sunbelt cities with low proportions. In part, it may be that, as new high tech complexes build their own professional/technical labor pools around them, few blacks are recruited because of their relatively low incidence in these occupations, but it could as well be a product of racial discrimination in selection of both sites and employees. At any rate, to the extent that plants dispersed in the 1970s, they evidently gravitated toward locations with low shares of black residents. This is particularly disturbing news for those who hope that the more standardized portions of high tech production, those which are presumably the subjects of dispersion, will help generate jobs for minority groups.

Disaggregating the model

By SMSA size classes

Some of the variables employed in the analysis just reported are rather closely associated with sheer size of cities. The presence of Fortune 500 headquarters and educational options, for instance, are features which are more or less related to the absolute level of population in a metropolitan area, having played a role in the growth dynamic of those complexes throughout their evolution. It would not be particularly helpful to smaller metropolitan areas to know that they needed a large educational establishment and a few more corporate headquarters really to find a place on the high tech map, if these are features that have some minimum scale that is not attainable by smaller communities. We therefore disaggregated the model into three metropolitan size groups, to test the strength of our variables in explaining within-group variation. The results should indicate which factors make cities more competitive with others of similar size in attracting high tech growth. Because the data set employed was so extensive, with 264 observations constituting a population, rather than a sample of metropolitan areas, it was possible for us to disaggregate by urban size categories with negligible losses in degrees of freedom. The results are described in Appendix 5.

A number of interesting contrasts emerge from the resulting regressions. It should be kept in mind that failure of certain variables to appear important in explaining within-subset variations does not mean that they are not crucial to high tech location. For instance, in none of the subsets were the amenities variables as important as they are in the overall regression results. Amenities may be critical in the location of high tech overall but unimportant in explaining the variation within sets of similarly sized SMSAs. Larger metropolitan areas may have greater amenities, which accounts for their relatively higher shares of high tech activity overall, but within-group variation amenities may be either insignificant or unimportant. Thus the absence of findings in any one set of regressions should not be taken as evidence that that variable is not important to the community or city involved. Rather, the evidence from the subset suggests that, to the extent communities are competing with others of a similar size range, their success will depend upon the factors significant in these regressions. On the other hand, a community might want to focus on attaining the threshold level of one of the variables which would help it transcend the group size boundary.

Overall, different locational factors do appear to be associated with variation among metropolitan areas in their size class than those associated with overall distribution. Climate, educational options, airport access and the presence of Fortune 500 headquarters were all significant factors in explaining overall metropolitan job change within subsets.[7] Fortune 500 turned out to be the only variable which was consistently related to high tech activity in all groups, and no single variable appeared to be statistically significant in explaining recent change across all three.

Paradoxically, in view of our expectations that agglomeration factors would prove more important in high tech growth among large metros and that labor force and socio-political variables would prove more significant for smaller metros, just the opposite was observed. Among the large set, the socio-political and unionization variables were significant in explaining mid-1970s job shifts while agglomeration factors were less powerful than they had been in shaping the overall distribution. This suggests that, to the extent that dispersion is taking place among major metropolitan areas, it is driven by the locational factors hypothesized to predominate in the more standardized functions of high tech activity.

In the medium-sized group, two agglomeration factors – business services and Fortune 500 corporations – maintained their significance in shaping mid-1970s growth differentials, in addition to the minority variable. For small cities, on the other hand, only business services could be confirmed as a significant factor in recent job change. Overall, then, agglomeration factors appear to be a route whereby smaller places can improve their high tech performance *vis-à-vis* each other, while "better" socio-political and labor force features are more apt to be associated with differential high tech gains among larger-sized cities.

It is important not to overstate the significance of these findings. For instance, although percent black could not be confirmed as a determinant of intrametropolitan growth differentials among small cities, it is a significant factor in shifting high tech growth from large to smaller places. Indeed, since the dispersion evidence suggests that intrametropolitan shifts from larger to smaller metropolitan areas are the most prominent form of high tech redistribution, the regression findings on the full set are more meaningful for policy purposes.

One particularly unexpected outcome of the model disaggregation is the finding that defense spending is a more significant determinant of differential high tech growth rates among large metropolitan areas than it is among small and medium-sized ones. There are two reasons why this might have occurred. First, the defense spending variable is constructed from data on prime contracts over $10,000. Geographers and regional economists studying the distribution of defense dollars have hypothesized that the absence of a good data series on subcontracting understates the degree of dispersion of these funds (Bolton 1966, Karaska 1967). However, analyses of a one-time-only set of data on subcontracting showed strong tendencies for subcontracting patterns to reinforce the extraordinary concentration of prime contracts awards, which are by and large channelled to major metropolitan areas and in a manner not very closely associated with population (Karaska 1967, Rees 1982, Malecki 1983).

Secondly, the 1970s were a period of defense build-down, particularly in the actual production of manufactured arms and materiel. Smaller cities, which were more apt to host factories for the more routine production activities in military-related high tech industries, may have felt the brunt of these cutbacks, while the more innovation-oriented activities, disproportionately centered in larger SMSAs, continued to be funded. Nevertheless, it is striking that these defense cutbacks were not also registered as negatively related to high tech growth differentials among the smaller-sized places in the period. Given the overall results, we must conclude that differentials in defense spending per capita are more important in explaining the differential growth shifts from large to small SMSAs, and among large SMSAs, than they are within the smaller city group.

By product–profit cycle industry groups

Grouping industries by their status in the profit cycle provided a second disaggregation of the locational model. *A priori*, we expected to find greater sensitivity to the agglomeration and amenities factors on the part of the innovating, rapid growth sectors. On the other hand, we expected to find greater explanatory power in the labor and access features in those high tech industries which are mature and/or declining.[8]

To test these hypotheses, we computed the high tech employment totals for 1977 for each SMSA for each industry group described in Chapter 4. We then ran regressions using the same high tech model on each group.[9] The results, contrasted with those for the entire set of high tech industries, are presented in Table 9.2.

Overall, we found that the model yielded a better fit for the innovating and expanding high tech sectors than it did for the volatile and more mature sectors. This confirmed the appropriateness of the model's specification for the sectors often referred to in the high tech development literature: electronics, communications, instruments, drugs and some types of chemicals, and selected machinery industries.

Surprisingly, the labor variables performed consistently poorly across all groups. Unionization rates for all five groups were either insignificant or perverse in their effect. Wage rates were positively related to 1977 job concentrations in declining high tech sectors but negatively related in the fastest growing sectors. Thus, low wages were more apt to matter to the leading edge industries than to older industries, contrary to our expectations. This may reflect the growing significance of labor costs over time, appearing particularly in the locational choices of the more newly established industrial sectors.

We found that the amenities and access variables generally retained their sign and significance across groups. However, educational options was the only variable consistently significant across the entire set. Freeway density was significant for all but the declining high tech sectors.

More striking was the finding that most of the other variables significant for high tech job patterns in the aggregate were not necessarily so for individual industry groups. Housing prices, for instance, were

Table 9.2 Location factors associated with high tech employment, 1977, by profit cycle group.

Locational variable	Hypothesized sign	All high tech industries	Profit cycle groups			
			Innovating	Expanding	Volatile	Mature	Declining
Labor							
wage rate	−		X				O
unionization rate	−	O		O			
unemployment rate	+			O			
Amenities							
climate index	+	X	X	X		X	
housing price	−	X		X		X	
educational options	+	X	X			X	X
Access features							
freeway density†	+	X	X	X	X	X	
airport access	+				X	X	X
Agglomeration							
Fortune 500†	+						
business services†	+	X	X	X		X	
university R&D†	+			X			
Socio-political							
defense spending†	+	X			X		X
percent black	−		X				O
R^2		0.52	0.47	0.45	0.39	0.40	0.29
F-statistic		20.7	16.8	16.0	12.2	12.9	7.9
Number of cases		264	264	264	264	264	264

X Significant in the hypothesized manner at 5% level.
O Perverse: significant in the *opposite* direction at the hypothesized level.
† Variable logged.

negatively associated with locational patterns in the expanding and mature groups, but not the innovative, volatile or declining ones. Climate was found to be significant only for the innovative and the mature categories.

Defense spending per capita, as expected, was significant for the volatile group. It was not significant for either the innovating or expanding sectors, although in each case the sign was positive. This suggests that the lure of defense spending on high tech may be confined to a number of heavily defense-dependent sectors, which are not necessarily to be found in conjunction with more steadily growing high tech activities.

Unexpectedly, both wage rates and racism appear to be more influential for explaining the location of jobs in the fastest growing, innovative sectors than for any other group. Airport access was more apt to be associated with employment distributions in the volatile and more mature high tech sectors than in the faster growing groups. This latter result may suggest that airport status lags new plant location, rather than acting as the incentive for plant siting.

In sum, the exercise suggests that agglomeration factors do appear to be more closely associated with faster growing high tech industry jobs. Also confirming our prior expectations, defense spending per capita was a more powerful factor in explaining the volatile military/industrial sector arrays. However, job distributions in the mature high tech sectors (TV and radio receivers, transmitters, plastics, inorganic and agricultural chemicals, paints and varnishes) were just as apt, if not more, to be significantly associated with amenities features as were the faster growing groups. Labor-related factors were not found to be more significant for the profit-squeezed industries. Access was only slightly more important to these same sectors than for their faster growing counterparts.

What do these disaggregated results look like from the perspective of each industry group? In the innovative fastest growing industries (the high tech headline catchers like electronics, computers, communications equipment, medical supplies, optical instruments, and measurement instruments), many of the popularly stressed locational factors do matter. Particularly, amenities like climate and educational options, and access to freeways and business services, were positively associated with high tech job incidence in 1977.

However, other hypothesized factors, such as housing cost, airport access, unionization and unemployment rates, per capita defense spending, and university-based research and development were not. And not all SMSAs with the favored features posted predicted job incidence in these industries. Among the ten worst outliers were eight SMSAs in the southern midcontinent from Colorado through Oklahoma, Texas, West Virginia, and North Carolina.

Among the expanding, modest growth sectors (like photographic equipment, drugs, construction and industrial machinery, organic chemicals, and engines), the locational pattern in 1977 appears to defy almost as many of the model's hypothesized relationships as it supports. Job distribution in these industries was positively associated with high wage and high unionization rates. Also perversely, it was negatively associated with university-based research and development spending. Apparently, these industries were historically neither attracted by defense spending nor strongly opposed to the presence of a black population. Nor has climate been a significant factor in their orientation. Of the hypothesized variables, only a few in the amenities, access and agglomeration categories appear to have had major influence.

Geographical job distribution in the volatile sectors (missiles, ordnance, aircraft, metalworking and electrical machinery, engineering instruments) appears to be statistically attributable to only four variables: defense spending, freeway density, airport access, and educational options. Neither high rates of unionization nor the absence of business services appears to have been a deterrent, while amenities are less important than they are for most other groups.

In the mature sectors (receivers, transmitters, plastics, paints, inorganic chemicals, etc.), amenities and access features, along with business services, appear to be more important in explaining historically evolved job distribution patterns than do labor and socioeconomic variables. Particularly surprising in this group was the failure of a single labor-related variable to operate in the expected fashion.

By far the poorest results, in overall fit and in hypothesized relationships, were found in the declining high tech industries (petroleum refining, railroad equipment, reclaimed rubber). Only educational options, airport access, and defense spending per capita figured significantly in this regression. No agglomeration factors appear to have been at work, as predicted, and neither do the labor market factors operate as we might expect in profit-squeezed industries.

The most interesting finding for this group is that the most extreme outlier category contains a number of manufacturing belt cities, whose 1977 high tech job totals were under-estimated by the model. These include the SMSAs of Johnstown and Erie, Pennsylvania; Gary–Hammond, Indiana; Bay City, Michigan; Youngstown–Warren,

Ohio; New Brunswick–Perth Amboy, New Jersey; and Huntington–Ashland, West Virginia/Kentucky. In most other groups, the same subset of SMSAs in the south and west comprised the worst outliers.

Summary

The results of the regression analysis indicate that the hypotheses imbedded in the model do explain a major portion of 1977 differentials in high tech job and plant location, both for the entire set of metropolitan areas and for the large and medium-sized subsets. They also performed well in explaining the changes in job and plant levels in the mid-1970s for larger SMSAs, less so for the successively smaller groups.

Our results indicate that the traditional labor supply characteristics, thought to attract industrial activity, are not very important in explaining the distribution of high tech industries at the metropolitan level. On the other hand, high tech activity is increasingly avoiding communities with large black populations – a disturbing phenomenon. Place features which are commonly thought to be important for attracting the professional segment of the workforce, including entrepreneurs, did figure prominently in the 1970s. In particular, educational options and climate appear to be strongly related to high tech location. Site features like freeway and airport access were more apt to be significantly related to plant location than to employment distribution; the proximity of agglomeration factors such as corporate headquarters and business services also appears more closely related to plant than job distribution. Defense spending, on the other hand, seems to be more closely associated with job than plant distribution.

We also looked at the changes in high tech location in the period 1972–7 to determine if certain factors were more powerful in explaining recent changes than the overall array of plants and jobs. Our results indicate that transportation access and agglomeration features were relatively less closely associated with the redistribution of plants and jobs than they were in the longer run, while amenities factors maintained more or less the same influence. The variables representing minority presence and per capita defense spending were even more important in explaining recent shifts in jobs and plants than in explaining overall locational patterns.

This is particularly true for the largest metropolitan areas. For smaller places, agglomeration factors such as corporate headquarters and business services continue to play a critical role in hosting high tech development, while access and amenities features are relatively unimportant. Overall, the four most consistently significant explanatory factors were educational options, business services, defense spending and percent black, suggesting an outline, at least, of the major distinguishing characteristics of successful high tech complexes.

The analysis, it must be stressed, merely confirms the existence of certain relationships; it does not determine the direction of causality. The regression results formally permit us only to reject the null hypotheses that the direction of the relationship is not the opposite of that hypothesized. We know, then, for our "determinants," only that they are statistically significantly associated with the endogenous variables. We cannot infer which variables are actually driving the location of high tech industries and which are merely accompanying it. It seems likely that the socio-political climate and access variables are truly causal in shaping locational shifts. The building up of a business services complex, however, may be more a companion to high tech development than a factor attracting it; the same may be true of some of the other amenities factors. In the case of a variable such as educational options, it is somewhat difficult to judge, since first-rate educational opportunities may both draw high tech around them but be dependent upon this type of activity for generating the interest, personnel and funding to improve educational offerings.

The implications of this analysis are that efforts to improve the accessibility of sites to both transportation networks and business service complexes are apt to be more effective than efforts to improve the "business climate" as registered in traditional factors such as wage rates, unionization levels, and a high cost of living. Cost factors in general appear to be less important than amenities, the availability of business services, and favorable receipts of defense spending. In the following chapter, when we come to draw out our policy implications, this will need to be kept in mind.

Notes

1   In preliminary regression runs we also included variables on presidential voting patterns and per student spending on elementary education (an amenity variable), but in no case were the parameters generated significant. We believe that these factors may play an important role in the siting of high tech activity *within* metropolitan areas.

2   In preliminary work, we also included the variable percent Latino population. This yielded insignificant results and was dropped in the final version.
3   In earlier versions of the model, we included an arts index which we later removed because it introduced too much multi-collinearity into the model. We also had an air quality index which performed perversely; eventually it was rejected because it turned out to include only measurable particulates from industrial services and not auto pollution.
4   In early runs, we also included a variable on industrial utility rates, as a proxy for access to and cost of energy. However, the results were not significant and the data were flawed by being available only for 1980 and only at the state level.
5   In preliminary work, we used a dummy variable which captured the highest ranked engineering and business schools in the country. However, this variable turned out to be too closely correlated with other metropolitan variables and inferior to the finer-grained data on university R&D.
6   It should be noted that these tests were performed before the era of airline deregulation, which has had a strong adverse impact on the dispersion of airport-dependent activities as the degree of monopoly pricing on intra-regional routes has increased while competition has driven down the price of interregional traffic, especially between major metropolitan areas.
7   Educational options and climate continued consistently to bear the correct sign, but could not be confirmed as statistically significant. Fortune 500 headquarters and airport access, on the other hand, assumed negative signs in several of the change regressions.
8   We did some preliminary work on individual industries, using an earlier version of the same model. The results were much less robust than for the aggregate set of high tech industries. However, a couple of interesting similarities and differences were noted. Fortune 500 was frequently (in 48 out of 100 industries) a significant factor in locational orientation, but quite often the relationship was strongly negative rather than positive. This suggests that some industries in our set are still quite heavily in their agglomerative stage, while the growth in others is primarily in the more standardized, dispersing functions. Of the variables which were frequently significant, airports bore the most consistently correct sign, in nine out of ten cases. Defense spending was the second most consistently performing variable. For a lengthier discussion of these results, see Glasmeier *et al.* (1983).
9   We experimented with truncated models for each group, based on initial regression results, using only significant variables. We found only modest change in the size of the coefficients, few instances of variation in significance, and no cases of sign reversal, despite the presence of moderate levels of multicollinearity.

10
What policy could do

Implications for national, state and local policy makers

THE MAJOR OBJECTIVE of this book has been to describe and to explain – not to prescribe. We will be satisfied, and we hope that most of our readers will be similarly satisfied, if the preceding chapters have thrown some new light on the nature of high tech industry, its patterns of growth and of organization, and its changing geographical distributions. But we are also conscious that many readers will have scanned these pages with another, urgent quest in mind. They are the congressmen, state legislators, local councillors and elected officials whose main concern is to understand how their local area is winning or losing out in the high tech stakes – and how, in the future, it might do better. To them, this final chapter is principally addressed.

Some policy-relevant findings

We start by culling, from the previous chapters, those conclusions that seem most relevant for policy formulation. In doing this, we want to stress the obvious fact that they all refer to past experience; they provide no guarantee that the future will be in any way similar.

High tech does create jobs

The first finding of the study with particular relevance for policy makers, we believe, comes in Chapter 3: it is that consistent with its popular image, high tech industry does indeed create large numbers of new jobs. Between 1972 and 1981 alone, the 100 high technology industries as defined for this study contributed a net growth of 1,080,000 new jobs: 87% of the entire employment growth in the manufacturing sector of the United States economy. Though it may be of small comfort to workers in those sectors and areas affected by the decline of older manufacturing industry, nevertheless at a time when the popular image is one of post-industrial service-led growth, with manufacturing in decline, this finding is a useful corrective.

Nor was this growth concentrated in just a few sectors. On the contrary, it was broadly based across many. True, only 7 of the 100 sectors contributed over 57% of 1972–81 high tech job growth. But 72 of the 100 showed some job expansion during the period 1972–81.

Some of the fastest growing high tech sectors in the late 1970s were in defense-related jobs: guided missiles, space-related aircraft and ordnance industries, computers, semiconductors and other electronic industries. And this was in a period when defense spending was only just beginning to recover from its post-Vietnam low point. Were we able to continue the analysis into the mid-1980s, we would be certain to discover a dramatic further increase. For policy makers at a national as well as a local level, unpalatable as the message may be to some, the lesson is clear: increases in the defense budget do spell high tech job creation. Of course, the converse is also true. Defense industries are among the more volatile high tech sectors; during the mid-1970s, some of them posted the sharpest job declines.

This underlines a point: although a big majority of high tech industries created extra jobs, a substantial minority – almost one-third of the total, over the 1972–81 period – recorded job losses. Some of these losses were large: over 33,000 in large arms ammunition, nearly 26,000 in consumer radio and TV, over 18,000 in synthetic fibers. So high tech, as such, is no guarantee of an expanding national or local economy: it has to be the right sort of high tech.

These declining high tech industries are a varied bunch. Some appear to have a saturated market, like alkalies and chlorine or reclaimed rubber. Some may reflect foreign competition: consumer radio and TV, phonograph records.[1] Product–profit cycle analysis suggests that many of these industries, though still classified as high tech in terms of occupational composition, were in fact far advanced in industrial life cycle. The conclusion seems inescapable: the product–profit cycle is not just an academic theory, but a fact which practical policy makers ignore at their peril.

High tech is on the move

The second policy-relevant finding comes in Chapter 6: it is that the majority of high tech industries show a distinct tendency to disperse. In the concluding words of the chapter: "there is much potentially mobile employment in high tech sectors, particularly those that are growing rapidly," which spells "good news for places that have to date missed out on high technology development." In particular, the dispersion rule applies strongly to the majority of the commercially oriented, more highly innovative industries that most people automatically associate with the high tech image. As these industries add jobs, these new jobs tend to locate in new places.

As usual, however, there are reservations. Chapter 7 shows that much of this dispersion takes a particular form: much of it consists of an outward movement from a few high tech core states into neighboring states (California to Arizona and Nevada; Massachusetts to Vermont and New Hampshire), and also of a shorter-distance movement from a few dominant high tech metropolitan areas into neighboring SMSAs (Los Angeles to Anaheim–Santa Ana–Garden Grove; San Francisco to San Jose; Boston to Worcester). In other words, it is relatively short-distance dispersal. And, with one or two exceptions (West Virginia, Ohio), much of the benefit goes to states and areas that have not traditionally been industrial centers.

The principal beneficiaries, indeed, have been in what can be called the defense perimeter of New England–Long Island, Florida and the West; the states and metropolitan areas that have conspicuously failed to benefit are clustered in the midwestern industrial heartland from Buffalo westward to St. Louis and Milwaukee. But the distinction is not the familiar Frostbelt/Sunbelt one. Old New England (Boston–Connecticut) and the Chesapeake/Delaware River valley region (New Jersey–Maryland) are both identified as high tech core regions, and the first has shown a tendency to dispersal that has benefited the neighboring states of Vermont and New Hampshire. The second, perhaps significantly, has not – thus providing an anomalous exception to the

High tech can thrive in old industrial regions

The New England example, just quoted, is the most dramatic illustration of the fact that high tech is not a monopoly of the Sunbelt states of the South and West. So are the Chesapeake/Delaware River area and Illinois. Three of the five major state agglomerations of high tech industry, with some 37% of the jobs in them, are thus in the nation's old industrial heartland. At the SMSA level, too, older urban agglomerations – Chicago, Boston–Lowell–Brockton–Lawrence–Haverhill, Philadelphia, Newark, Detroit, New York, Cleveland – actually emerge as among the biggest concentrations of high tech employment. And two of these – Boston and Worcester – were among those ten SMSAs posting the largest job gains in the mid-1970s.

Admittedly, there were only two such; they were dominated in the listings of job growth by the Sunbelt SMSAs like San Jose, Anaheim, Houston and San Diego. And in terms of percentage gains, the records were held by small, fairly isolated urban areas in the interior of the United States, far distant from the major metropolitan areas. Nevertheless, the example of New England does illustrate convincingly that there is no fatalistic rule that the march of high tech is southward and westward away from the old industrial core. Much, still, depends on the ability of an old industrial region – here, literally the nation's oldest – to reconstruct itself. And this ability will turn in large measure on the region's inheritance of physical and social infrastructure and accumulated business services: the factors that are examined in Chapters 8 and 9 of the study.

Small places can win in the high tech stakes

The metropolitan areas that posted the biggest percentage high tech gains, whether in new plants or in new jobs, are overwhelmingly small – even, it dare be said, obscure – places. They are the beneficiaries of the deconcentration of high tech from its original core concentrations, even though many may have benefited only through a single branch plant. Most readers would not readily identify on a map such centers as Lawton, Oklahoma; St. Cloud, Minnesota; Laredo, Lubbock, or McAllen–Pharr–Edinburg, Texas; Fort Meyers, Panama City or

Lakeland, Florida; or Cedar Rapids, Iowa. Yet all these were among the most dynamic high tech gainers in the United States, during the 1972–7 period, in terms of their percentage gains in high tech plants or jobs.[2]

Chapter 7 points out one intriguing feature these places have in common: they are almost all located in the interior of the country, and most are far-distant from the major high tech core centers. It may be that they are benefiting from branch plants or spinoffs from the largest centers; it may be that they are centers of defense-based high tech. What is certain is that they lack the inducements that have traditionally been thought important for attracting high tech: major research universities or an array of other educational institutions, a well-developed infrastructure of business services, or concentrations of major industrial headquarters. For the routine high tech activities that are the basis of growth in so many of these places, such factors may be irrelevant. If they can succeed in the high tech stakes, then, presumably, so can many other such places.

The critical question, however, is whether such small-town high tech establishments will eventually bring integrated, self-sustaining local economic growth. Recent research (Glasmeier 1986) is pessimistic on this score: it suggests that most of them concentrate on the least technical processes, which are least likely to engender spinoff. Even the select minority that seem to be developing higher-level high tech – places like Austin, Texas; Portland, Oregon; and Melbourne, Florida – are by no means replicating the story of Silicon Valley or Route 128. Rather, their development is tied to existing high tech centers through the medium of technical branch plants which will not necessarily generate integrated growth. Such growth, it seems, depends on a complex set of forces associated with the nature of the product and of the productive process, corporate attitudes toward local procurement of inputs, and corporate policies on spinoffs. And such factors have not historically been open to conscious policy intervention. So high tech growth in small places should perhaps be welcomed realistically for what it is: a useful source of local jobs rather than a potential source of indigenous, integrated growth.

High tech location is explicable, therefore manipulable

This, for policy, may be our most important finding. Chapter 9 has shown conclusively that the long-term location of high tech industry – and, to a lesser extent, short-term changes in location – can be explained in terms of a few key location variables. These variables, for the most part, are the ones often surmised to be important for high tech industry: they fall under the three headings of amenities, access, agglomeration. Amenity variables – a good climate, a range of educational options – are particularly significant for long-term location. Access to the interstate highway system, and to a good airport, are also important, and there is a suggestion that airport access may be a critical variable for small places aspiring to high tech status. Agglomeration economies – the presence of major headquarters companies and a wide range of business services – are a further significant feature. For shorter-term changes during the 1970s, amenity variables remained strong; access variables diminished in importance, and the effect of agglomeration variables was more mixed, although business services remained significant. Above all, during this period, the importance of defense spending increased.

Our analysis was equally striking when it came to the factors that were insignificant, or negative, in their effect: unionization, wage rates, and – most surprising of all – a strong research component. The last proved to have little significance; indeed, it was negatively associated with high tech growth. And, disturbingly, it appears that high tech avoids places with strong concentrations of black population – although, as so often in statistical analysis, this may prove to be a measure of some other factor.

For policy makers in the scores of smaller metropolitan areas, particular interest may lie in our separate regression analysis for such places. Here, the long-term pattern suggests that the key factors are access to the interstate system, a business service center (preferably with a major headquarters office), and significant defense spending. In comparison, other factors – low wages, weak unionization, good climate, cheap housing and good educational options – would not provide the critical attraction. But, disappointingly, our analysis of recent changes in these smaller places gave very poor results. As a result, we would be reluctant to base any policy prescriptions upon them.

There is another group of policy makers who may look with particular concern at this analysis: those in the bigger, older cities that have been so disproportionately affected by the decline of the smokestack industries. Here, our most significant finding is that traditional agglomeration economies seem to be losing their significance. Amenity variables, too, are of little importance – a disappointing result, perhaps, for those who place faith in the superior cultural resources of the big cities. Most significant of all are the findings for recent big city change: defense spending has a very significant positive effect, while the presence of minority population has an equally significant negative effect. What this seems to be saying is that older major cities with big minority populations are doing badly in the high tech race, while newer big cities with a strong basis of defense industries are doing well.

Implications for policy

In drawing policy implications from these findings, we have to be extra tentative. Generalizations necessarily have many exceptions; and we are very conscious that we have not disaggregated our analysis nearly as finely as we would have liked.

Implications for local policy

That said, the following conclusions seem to emerge for local policy makers at municipal and state level.

Almost any place can compete   The analysis strongly suggests that indeed some factors are negative, others strongly positive. But – providing macro-economic conditions, in the form of general economic growth, are right – hardly any place scores so badly that it is outside the race. With the possible exception of some of the older, isolated, single-smokestack-industry centers, almost any state and city can work to strengthen its positive, high tech attracting factors. It can improve its accessibility to the national highway system; improve its airport; work to strengthen its attraction to business services of all kinds, including headquarters companies. It can simultaneously work to improve its amenities. This applies both to large cities and to small, though the precise strategy will vary from one to the other.

True, such developments cannot occur overnight. Remote places cannot suddenly become accessible; a small-town airport cannot become a Chicago or an Atlanta; a place cannot change its climate, though it can improve other amenities. The foundations for successful high tech growth have to be built, and that will take time. Nor will development, when it comes, echo the story of places like Silicon Valley or Highway 128, for it is all too likely to remain dependent on headquarters offices in such places. Nevertheless, it can provide local jobs.

Defense now the key factor   The analysis is overwhelmingly clear and unambiguous on one point: during the 1970s, defense spending emerged as one of the key variables explaining high tech growth. Though we lack direct evidence for the later period, we are certain that if anything this conclusion would be strengthened. Though high tech development broke away from its defense origins during the early 1970s, it subsequently returned to them – to a greater degree, perhaps, than at any time since the 1950s. The present proposals for the Strategic Defense Initiative ("Star Wars"), with their $26 billion research program over five years, underline this point. Described as the biggest research program in history, the Strategic Defense Initiative (SDI) seems certain – unless halted – to bring an enormous military procurement program in its wake. And this, we can be certain, will powerfully shape the pattern of American economic growth during the 1990s and beyond.

How it will do so is not precisely clear. The regional and local economic impacts of the defense program have not, we believe, been explored in sufficient detail. As we close this study, we are starting another to examine precisely this question: how far and in what ways did the expansion of defense spending, in the decades following World War II, reshape the industrial map of America? If we can answer that question, we shall be better placed to predict the ways in which it might do so again in the decades to come.

The missing variable: research The most surprising conclusion of this study, for us and doubtless for most readers, is that research spending did not prove a significant factor in explaining the long-term location patterns, or the short-term shifts, of high tech industry.

Possibly, if we could somehow disaggregate the research spending variable in our analysis, we could provide a better explanation of what otherwise remains an enigma. Otherwise, one of the most cherished myths of high tech policy – that a strong research university is the key to high tech growth – seems to be without empirical foundation. We are inclined to think that the explanation lies in the importance of the defense variable: what was earlier thought to represent the importance of fundamental research, as in the growth of high tech along Highway 128 around Harvard and MIT, or in Silicon Valley next to Stanford, really represented very highly concentrated defense spending in one or two key universities. An alternative explanation is that these universities happened to have just the required research strengths in electronic engineering at the critical time. Universities that developed the appropriate programs later, like most of the big midwestern centers, then simply found that they were exporting their graduates to the high tech cores. Thus, whether the origin lies in defense or not, it is the character of existing research at a point of time that proves vital.

Will the future replicate the past? Nothing is more fatal, in social science research, than to draw certain statistical conclusions from past data and to project them forward into the future. Our regression analysis provides good proof of that. The relationships that appear significant in 1977 – which, we assume, represent the cumulative impact of development over a long period – are by no means the same as those that emerged as important during the short period 1972–7. Doubtless they are already different again. Key changes could occur, in the coming years, both on the supply and the demand sides of the high tech equation.

On the supply side, product innovations could create yet new high tech industries, with quite different originating geographical patterns than the ones of the recent past. On the demand side, shifts in federal policy could lead to a massive fall in defense spending and a corresponding rise in other, peaceable high tech industries. As the Japanese example so amply shows, eminence in high technology does not have to come from a military base. One element of a national strategy would certainly involve the improvement of product engineering so as to restore America's former pre-eminence in capital goods and machine tools. This would particularly help industrially depressed Midwestern centers such as Chicago, Cleveland, and Milwaukee.

On the other hand, there is a considerable inertia in the system. Especially if changes in the high technology mix came through government procurement, it seems likely that the new high tech applications might be produced in the same locations as the old. Peaceable exploration and exploitation of space, for instance, would probably utilize the same kinds of technologies as the SDI program, and would doubtless go to the same contractors. Likewise, peaceable uses of biotechnology would be developed in the same laboratories as now do research in biological warfare. The only exception to this rule would come if genuinely new supply-side innovations emerged, like wild cards, in places remote from existing high tech centers. And this, by definition, is unpredictable. The safe strategy, therefore, would be to build on the existing high tech strengths of the various core regions (as outlined in Ch. 7) by seeking to develop new and related high tech activities there.

Implications for national policy

One fact that emerges clearly enough from our analysis is that high technology development at local level (or the lack of it) depends very much on policy decisions at the national level. And there, it is unrealistic to assume the development of a bipartisan policy. In the future, as during the past 30 years, the federal influence over local areas – the pattern of disbursements, direct and indirect, into defense, research and development, transportation and communication infrastructure – will reflect both ideology and political obligation. So there could not be one set of implications, but two.

A Republican scenario A Republican administration in Washington, we would surmise, would follow very much the policies of the early 1980s – which happen to be the same, essentially, as those followed by previous Republican administrations (Mollenkopf 1983). Defense spending would be increased. Large contracts would go to high technology producers in such areas as aircraft, missiles, and associated electronic equipment. The bulk of these contracts would clearly go to existing producers in existing centers in the areas we have designated high tech core regions, especially in the West and South. There would be very little or no attempt to follow any kind of regional or urban redistributive policy in order to divert any such funds into the distressed cities of the old manufacturing belt. This, we surmise, is the policy now being followed and intensified under the SDI program. It will further boost the economies and the fortunes of such major high tech concentrations as Greater Los Angeles, the San Francisco Bay Area and southern Texas; but it is likely to accelerate the process of further outward spread from them into neighboring states in the South and West.

A companion piece is the escalating role of arms in the US trade balance. While net exports of non-military capital goods declined by $30 billion between 1980 and 1984, military exports grew by $11 billion, up 200%. These sales, facilitated by the Pentagon and by tied aid policies, enhance the growth of the same regions that are producing military hardware for the Department of Defense.

The regional result of a continuing Republican policy will be the further boosting of the "defense perimeter:" high tech concentrations from New England through Long Island, eastern Florida, Texas, California, and Seattle. The fastest growing cities will continue to be the medium-sized, heavily defense-dependent places like Colorado Springs, central Utah, and Huntsville, Alabama, while the high tech peripheries of agglomerations like Los Angeles, Boston and Silicon Valley will receive the overflow. This dramatic recomposition of domestic and internationally oriented manufacturing has been the major factor behind the extraordinarily uneven regional growth rates of the first half of the 1980s – the Sunbelt captured fully 91% of net population growth. Engineers and technical people will continue to be siphoned off from midwestern universities to Sunbelt high tech complexes.

Two Democratic scenarios It is more difficult to imagine what high tech futures the USA might have under a Democratic administration. Two camps appear to be vying for leadership in shaping platform priorities. Both would undoubtedly argue for a leaner defense budget. But their respective visions of the purposes and direction of high tech are quite starkly different.

One group stresses the engenderment of "sunrise" industrial sectors in the US economy.[3] Their view is that it is imperative that the USA compete successfully with its industrialized competitors, especially Japan, who have used targeted intervention to ensure that certain key industries become efficient, high quality producers that can flourish in international markets. They argue that a set of key areas – biotechnology for agricultural and medical applications, peaceful space exploration, the fifth generation computers – should be designated for joint public–private investment along the lines of Japan's highly successful program.

Policy tools suggested for this effort include the provision of research dollars for basic industrial innovation (much as the space and defense programs have done for the aerospace, electronics, and instruments sectors). There would be a national development bank to supply capital in the form of low interest loans and equity shares for industrial revitalization investments. Aid to state and local governments would help them gear up to compete, by providing earmarked funds for science parks, linked with universities, and for the development of transportation, communication, infrastructure and educational services. State and local governments would be encouraged to provide venture capital funds, product development grants, site subsidies, and retraining programs to help new entrepreneurs compete and/or to lure new branch plants.

This type of program would undoubtedly smooth out the interregional growth process, stemming the extraordinary shifts produced by defense-related high tech booms. It would help revitalize basic industry in the nation's manufacturing heartland, and restore to some degree diversity and strength in our export trade. It would slow down the technical brain drain from the Midwest.

Critics of this formulation[4] go deeper into the issue of "innovation for what?" They take exception to the challenge "Can we compete?" and ask, "What's the game, anyway?" In their view, innovation ought to be directed toward the basic needs of the society, such as nutrition, housing, transportation, health, and infrastructure. Investment in innovation should be directly linked to a public consensus on what breakthroughs are needed. Priorities like better birth control, cheaper and speedier mass transit, better quality housing at affordable prices, and cleaner water, air, and neighborhoods would create a very different set of research and development spending patterns than do those dictated by "the market."

Furthermore, this left Democratic position would explicitly calculate into its high tech policy the cost to society of worker displacement due to automation and of community destruction in the process of remaking the urban and regional environments. "Growth for communities" would permit the targeting of public incentives to encourage high tech producers to take advantage of existing labor skills and urban infrastructure, rather than taxpayers subsidizing the migration of technical workers and engineers away from existing centers and the building of new cities, highways, sewers, and water systems in newly created high tech centers. This approach would significantly cut the social costs, borne by both individuals and taxpayers at all levels of government, of the excessive shifting now taking place in high tech related production and settlement patterns.

All three of these scenarios are provocative. They exhibit large differences in vision, priorities, and prescriptions among them. These underscore the seriousness and challenge of the choices ahead. Innovation and high tech production are not going to disappear – but their nature, and the kinds of communities they create, remain central questions on the US development agenda.

Notes

1 In the period after 1977, the phonograph record industry showed a distinct recovery.
2 It should be stressed that *absolute* employment gains in such places were small.
3 Examples of this line of reasoning include Lawrence (1984) and Thurow (1980).
4 Examples of this position include Bowles *et al.* (1983), Kuttner (1984), and Bluestone & Harrison (1982).

APPENDIX 1

Sources of occupational data: Census versus OES

THERE ARE TWO comprehensive sources of national industry occupation data: the 1970 Census-based industry occupation matrix and the 1980 Occupational Employment Statistics (OES) survey-based matrix. The OES survey is a federal–state cooperative effort undertaken on a three-year cycle. States receive the technical specifications for the survey from the Labor Department. After the completion of a full three-year cycle the manufacturing data are forwarded to the Department of Labor, which then compiles the data into the Industry-Occupational Employment matrix.

For a number of reasons, the OES-based matrix proves to be a more reliable and comprehensive source of occupational data than the Census. First, the Census is a survey of individuals, whereas the OES survey is a survey of employers. In the Census, individuals self-select the occupational category they consider most closely fitting a description of their job, with little or no capacity for listing multiple occupations if the individual holds more than one job. The OES survey, on the other hand, is a survey of jobs within establishments and thus avoids the problem associated with secondary employment.

Secondly, the OES uses a more precise and careful schedule of occupations definitions based on skill levels. In the Census, occupational titles form a category which often includes individuals with widely differing skill levels. Similar occupational titles used in the Census and the OES are not necessarily comparable because the Census titles usually take the name of the most prominent occupation within that group. Also, the OES reports individuals performing one or more jobs on the basis of the job requiring the highest skill level. The Census reports individuals on the basis of the occupation in which the greatest number of working hours are spent.

Appendices

Coding procedures are substantially more accurate for the OES survey. The OES survey is filled in by an official within the firm whereas the Census is filled in usually by one individual on behalf of the household. Another source of coding error in the Census stems from the lack of sufficiently detailed occupational categories. National Science Foundation estimates that the range of error in the Census survey is as high as 50%. The OES survey questionnaires are tailored to specific industries. Each questionnaire includes a maximum of 200 occupations; residual categories are used so that establishments can report total employment. Employers are requested to specify newly emerging occupations if they are not already found on the questionnaire.

Finally, the industry disaggregation of the OES is superior to that in the Census. The Census covers 201 industries at varying levels of two-, three-, and four-digit disaggregation, based on the 1967 SIC categories. The OES uses the 1972 three-digit SIC classification for 378 industries. In the Census, inclusion of four-digit industries is based solely on response rates, whereas the OES is systematic in reporting all industries at the three-digit level. The Census reports 377 occupations while the OES reports 1,678 occupational categories. For these reasons the OES provides more systematic and detailed descriptions of individual industry occupational profiles.

Identifying high technology industries by occupational mix

We explored two alternative measures of occupational mix to identify high tech industries at the national level. We could not use all professional and technical workers, as this includes many occupations such as teachers and social workers not clearly related to high tech output. The first used the proportion of three occupations (engineers, engineering technicians, and computer scientists) as a percentage of total industry occupations. There were 27 three-digit industries which exceeded the manufacturing average of 5.51% of engineers, engineering technicians, and computer scientists.

In a second alternative, we added two additional occupational categories, life scientists (including chemists, geologists, physicists, and biological scientists) and mathematicians. We believe that many of these employees are engaged in product conceptualization and development resulting in high tech production. To exclude them might be to fail to capture activities like biotechnology and new materials. They were added to see whether their inclusion significantly changed the identification of industries or altered the original rank ordering. The results are shown in the seventh column of Table 2.5, their ranking in column 1. We found that their inclusion did not significantly change the rank ordering but increased modestly the number of industries designated as high tech. Two industries, reclaimed rubber (SIC 303) and paints (SIC 285), moved up in rank, to 22 from 28 and to 21 from 30 respectively. Other minor changes in ranking occurred as well. Including life scientists and mathematicians enlarged the pool of high tech industries to include SIC 384, 303, 287, 285, and 284, while eliminating two other industries, SIC 369 and 344, thus increasing the total number of industries to 29.

We decided to use the second, somewhat broader, definition of high tech occupational mix. We also chose a relatively generous cutoff point: all sectors exceeding the manufacturing average. We could have adopted a more restrictive cutoff point, such as a professional/technical occupational share in excess of 50% of the manufacturing average. However, there seemed to be no theoretical rationale for doing so. Opting for a fuller definition at the outset permitted analysis at a more general level, with more contrasts among the sectors examined, without precluding the study of a more limited set at a later stage. It also highlighted industries with innovation potential.

We also considered using as our cutoff point the durable manufacturing average. We did so because we wished to compare our results with a study by the Massachusetts Manpower Development Department which used that measure.[1] This reduced the pool of our industries to 21, with the deletion of SICs 284, 287, 303, 354, 356, 365, 374, and 384. Sectors eliminated by this procedure included agricultural chemicals, metalworking machinery, and medical equipment industries, all of which have substantial high tech components. The seven rejected industries would diminish the high tech employment base by 26%. We decided it would be better to include rather than exclude them, since their presence is so substantial, and since there is no theoretical reason why a "durables only" definition would be preferable.

Our final definition, then, consists of those three-digit manufacturing sectors whose proportions of engineers, engineering technicians, computer scientists, life scientists, and mathematicians exceed the national average for manufacturing industry; they are listed in descending order in Table 2.5.

Two final comments need to be made. First, we chose to limit ourselves to the manufacturing sectors because we could not procure spatially disaggregated data of similar quality for other sectors, particularly services. This is a clear limitation, since it is generally acknowledged that sectors such as computer software and commercial R&D labs, which fall into the business services category, deserve to be regarded as high tech. Fortunately, the number of such sectors is believed to be small (see OTA 1983). In addition, many of these sectors are geographically linked to manufacturing sectors as members of larger high tech growth complexes, in which the key location decisions are made by the latter. In our previous work on the computer software sector, for instance, we found it heavily clustered around computer design and manufacturing sites (Hall *et al.* 1985). We believe that the inclusion of these sectors would not seriously change our aggregate findings on the location of high tech industry in the USA although it might enhance the lead role ascribed to new centers of high tech growth such as Silicon Valley.

Secondly, we have defined high tech at the three-digit level. We were unable to do so at the four-digit level because the occupational data with that degree of disaggregation does not exist.[2] This means that we cannot tell whether all four-digit components of a three-digit industry would be classified as high tech, lacking a true picture of the occupational composition of each. We could have chosen under these circumstances to do *all* of our analysis at a three-digit level. Yet this would have only obscured, not eliminated, the problem. We chose to pursue the locational analysis at the four-digit level believing that our audience would prefer to see the detail and could thus more easily draw their own conclusions about the appropriateness of the classification in each four-digit industry's case.

Notes

1 Our comparison showed interestingly that Massachusetts' industry occupational configuration does not mirror the nation's. Their results using both the state manufacturing average of 8.7% for technical occupations and the durable average of 13.7% differed substantially from ours. Massachusetts has a high proportion of high tech industries in general and hence a high overall proportion of technical occupations represented in the workforce. They justified their use of the durable as opposed to the manufacturing average on the basis of industries represented in the state. Yet no conceptual reason for this choice is offered (Vinson & Harrington 1979).

2 No other research effort that we know of has been able to do a systematic four-digit delineation.

APPENDIX 2

The Census of Manufactures plant location data base

IN ORDER TO test our hypotheses about high tech agglomeration and dispersal, we assembled an extraordinarily fine-grained data set. This set contains plant and employment estimates for each of our 100 industries for all 3,140 counties in the USA. What follows is a brief description of this data set, which is relied upon for most of the empirical analysis in Chapters 7 and 9, along with some notes on data and estimation problems.

The *Census of Manufactures, location of manufacturing plants* tapes offer the most detailed plant-level data for manufacturing published by the federal government. For every county, the tapes enumerate all plants in each of the 450 four-digit (SIC) manufacturing industries. The tapes do not include actual employment levels. Instead, the data are arrayed in seven employment size classes for all counties in the United States. We used this data on plant incidence to estimate employment levels using national four-digit industry employment means.

The data base and previous users

The disclosure restraints on federally collected business data have long plagued geographers and economists working with regional analysis of industries. Published Census reports contain data on the number of establishments, aggregate employment, payrolls, value added, value of shipments, and other items. While data are also published at an area level (states, SMSAs, counties, and major cities), they are not complete when their publication would disclose information on individual companies. This results in very uneven coverage at every spatial scale below the state, especially for three- and four-digit sectors.

The *Location of manufacturing plants* tapes offer a route out of this data morass. They are machine-readable files giving the number of manufacturing establishments in each of the approximately 450 four-digit manufacturing industries, grouped by employment size class, and identifying the state and county within which they are located. Each record is associated with a particular industry (four-digit SIC) and a particular county (with a state or US summary) and contains the number of establishments in seven employment size categories: 1–19 employees, 20–49, 50–99, 100–249, 250–499, 500–999 and 1,000 employees or more. There is a record for each industry–county and industry–state combination with one or more manufacturing plants.

The employment size class assigned to each plant is based on the average total employment reported by plant. Both production workers and "all other" employees at the plant are included in this total figure. The data in each Census year, 1972 and 1977, correspond to contemporary county definitions.[1]

The location data tapes provide no information on actual employment levels other than listing plants within employment size classes. Changes in the array of plants are themselves interesting. But the employment size of individual plants could vary dramatically across regions and thus are not a reliable proxy for job distribution. Indeed, theoretically, we might expect dispersed branch plants to be larger than small innovative plants, though smaller than headquarters facilities. To overcome this problem, employment estimates were constructed by calculating the mean for individual industries using plant level information from the tapes and published national industry employment data.

Few researchers have attempted to estimate plant-level employment levels from the location tapes. This may be due to the fact that there are no complete industry-wide four-digit employment means data with which to construct employment estimates. It may also be related to the fact that the data are available only in machine-readable form. Staff in the Industry division of the Census indicated that they were unaware of attempts to develop employment estimates from the tapes. However, one researcher, Michael Conroy, at the University of Texas in Austin, estimated county employment from the 1972 location tape (Conroy 1975).

Conroy used the national two-digit manufacturing means for the six employment size class intervals. Conroy indicated that his estimates were not accurate for predicting specific plant-level employment. He found, however, that employment estimates at the multi-county, regional, and national levels were considerably more accurate and reliable, and correlated well with published employment figures.[2] Conroy argued, in fact, that the two-digit means were the maximum likelihood estimate of the number of employees in a plant in a particular size class from a randomly selected county and industry. He further suggested that the estimation procedure could be substantially strengthened if more disaggregated industry means were made available. At the time of his research, no finer disaggregation was available.

Estimating the means for "high technology" industries

Our approach to estimating employment means is far more detailed and disaggregated than previous attempts. Instead of using two-digit published means, we calculated them on the basis of a combination of four-, three-, and two-digit national industry employment data. In this procedure two data sources were used. Table 4 of the *Census of Manufactures, industry statistics* for 1972 and 1977 contains national data on the number of plants and actual employment levels for four-digit industries. *County Business Patterns*, Table IB, for 1972, 1974, and 1977, contains data similar to that found in the Census, but is based on a smaller sample size. In addition, the reporting procedures used in *County Business Patterns* publications are less precise than those of the Census. However, use of the latter was necessary as a "second best" because in approximately 26% of the industry employment size classes, data were missing in the national four-digit Census of Manufacturing published industry reports.

Missing data consisted primarily of two types. The first occurred in the lower end of the employment distribution in size categories 1–4, 5–9, and 10–19. In this case, the number of plants were listed but the missing employment figures were identified by the symbol Z indicating the category contained less than 50 employees. In the majority of these cases, only one size class was missing data. Since we were collapsing the first three categories to match the location data tape employment size class 1–19, the effect of this missing data was not significant in most cases; for example, fewer than 23% of all missing data points were in size class 1–19.

The second type of missing data occurred in the middle and upper end of the intervals where there were only a few reporting units. In this case the number of plants were reported but employment was not disclosed. Rather, the employment data were collapsed into the previous size category. Fully 31% of the cases of missing data occurred in the largest employment interval, 2,500+. Since we collapsed the last two size classes into an interval of 1,000+ to match the location tape, this missing data presented minor problems except where more than one of the largest employment classes were empty. For 1972, 7% of all industries were missing data in two or more size classes in the upper range; for 1977, 8% of all industries were missing data in two or more size classes in the upper range. In these exceptional cases we calculated the means on the basis of progressively less disaggregated data.

In all cases we attempted first to calculate industry means on the basis of Census of Manufactures four-digit published data. Published Census data were available for approximately 80% of the industries' employment size classes. In cases where Census four-digit data were not available, a second procedure was applied. This procedure differed slightly for 1972 and 1977. For 1972, where four-digit data were not available from the Census of Manufactures, we calculated means based on the two-digit Census of Manufactures published data. When two-digit data were also missing we used three- and four-digit data calculated from *County Business Patterns (CBP)*. The use of 1972 *(CBP)* data posed serious problems because, whereas the Census of Manufactures published data reflected changes made in the Standard Industrial Classification Manual (SIC) in 1972, *County Business Patterns* did not correct for these changes until 1974. Because we were particularly concerned about many of the industries that were changed in the SIC revision, we chose to use 1974 *(CBP)* three- and four-digit data rather than the uncorrected 1972 *(CBP)* version.

We recognized that certain industries' employment could have changed between 1972 and 1974, but there was no conclusive way to test for such differences. As Table A2.1 indicates, 87% of the missing four-digit data for 1972 was supplemented by using 1972 Census of Manufactures two-digit data, while 13% were estimated using *CBP* data for 1974.

For 1977 a slightly different procedure was used. In cases where four-digit Census data were missing, we calculated means based on 1977 four-digit *CBP* data. Again, as with the 1972 data, when four-digit data were missing, first three-digit *(CBP)* data were used, and when data were not reported at the three-digit level, we used two-digit Census of Manufactures data. As Table A2.1 illustrates, for 1977, between 98% and 99% of the missing Census data was supplemented with four- and three-digit data from County Business Patterns.

Table A2.1 Data sources used to calculate missing data in employment.

1972
In 1972 23% of intervals had missing data.
159 missing values were estimated of which:
    138 or 87% taken from two-digit Census of Manufacturers 1972
      16 or 10% taken from three-digit County Business Patterns 1974
       5 or 3% taken from four-digit County Business Patterns 1974

1977
In 1977, 19% of all intervals had missing data.
131 missing values were estimated, of which:
    115 or 88% were taken from three-digit County Business Patterns 1977
     14 or 11% were taken from four-digit County Business Patterns 1977
      2 or 2% were taken from two-digit Census of Manufactures 1977

Test of mean estimates

In order to test for systematic error we ran regressions on the estimated industry employment distributions. First, regressions were run for a number of industries with reported four-digit means to examine the relationship between the expected and actual means. This procedure was then applied to industry size classes with missing data.

In the majority of cases where missing data occurred at the lower end of the distribution, the R^2 for the estimated means was greater than 0.97. When regressions were calculated for cases with missing data, in the upper end of the distribution, the R^2 varied between 0.86 and 0.97. It became apparent that, given the data were arrayed in set intervals, the relationships were bound to be linear. Thus deviations from the interval means were not expected to be large. We therefore concluded that calculating means on the basis of sequentially less disaggregated published industry data was sufficient for our research effort.

Cautions on the interpretation of employment data

The reader should keep in mind throughout the locational analyses that while all figures on plants are actual Census figures, and therefore highly precise at the county level, the job numbers have been estimated with the procedure just described. We are quite confident of the quality of these estimates for individual industries at the regional and larger metropolitan levels of aggregation. For almost all counties, we are confident that the aggregate number of estimated high tech jobs are close to the actual populations.

However, for some individual counties, especially those with few high tech plants, it is possible that our estimates may seriously understate or overstate jobs. This will happen only if there are one or more plants which fall in the open-ended interval of 1,000+. To take an outlying hypothetical example, suppose a rural county has one enormous missile plant which actually employs 9,000 workers; our data base will accurately reflect the one plant but will under-estimate employment at 5,313, the national mean employment level for that interval. Similarly, high tech jobs in a county with a single 1,005-person missile plant recorded in the 1,000+ category will be over-estimated at 5,313. In reality, however, there are very few counties in the USA which would pose this problem. Most counties with large high tech plants have large numbers of plants and jobs in other industries and size intervals as well, minimizing the magnitude of the error of the estimates. Nevertheless, we do urge caution in interpreting the results for smaller units of spatial aggregation.

A second caution is that our data base does not distinguish within industries between plants with a preponderance of high tech occupations and activities and those with very low levels. It may be, therefore, that a semiconductor plant in Arizona or Colorado is a relatively routine operation with low levels of professional/technical labor and would not qualify on a *plant* basis as "high tech." But since no data is available on the occupational structure of industries on a plant by plant basis, we have had to treat all plants in SIC 3674, semiconductors, as high tech. Thus some counties will appear to be more "high tech" than they merit, and this bias will generally work against the centers of innovation and in favor of outlying areas which are the recipients of the more routine functions. On the other hand, this is not an unmitigated pity, since many advocates of sunrise industries are interested in precisely this kind of diversified job generation potential. The reader should simply keep in mind that we are tracking the distribution of plants identified as belonging to high tech industries, not necessarily plants which are themselves engaged in sophisticated innovative production processes.

Notes

1   In order to ensure comparability, we had to account for discrepancies in data collection and reporting between the two Census years. There are two notable differences between the 1972 and 1977 tapes. First, between 1972

and 1977, two minor revisions were made in the Standard Industry Classification (SIC) manual. The 1977 tape reflects these changes. In 1977, a new industry, motor homes (3716), was separated out of truck and bus bodies (SIC 3713). As SIC 371 did not qualify as a high tech sector, this change did not affect our work. A second code change was the grouping together of SICs 3671 (TV and radio receiving tubes), 3672 (cathode ray TV tubes), and 3673 (transmitting and special purpose tubes) in 1977 to form industry 3671 (cathode ray tubes, NEC). In this case, we chose to use the 1977 definition, aggregating the earlier three industries for cross-temporal comparisons.

2 Although the percent error for individual regions ranged between 36.5 and −14.1, Conroy argued that his results were fairly sound and unbiased when aggregated across all regions. Readers are referred to Conroy (1975) for additional detail on tests of the employment estimates.

APPENDIX 3

The independent variables

Airport access

AN INDEX OF accessibility which allocates a rating from 0 to 4 to SMSAs with the following designations: no airport, non-hub, small hub, medium hub and large hub. The numerical values assigned to the index are based on the percentage of US passenger traffic handled. The values for the scaled increments are as follows: 0, no airport; 1, less than 0.05% of total US passengers served; 2, 0.05–0.24% of total US passengers served; 3, 0.25–0.99% passengers served; 4, 1% or more carried from the airport in 1978–9.

It is important to note that some areas, adjacent to large metropolitan areas such as New York, do not have their own airport facilities but rather rely on the adjacent international airport services available. We considered assigning the same designation to these places without airport service as their adjacent SMSAs. Areas which were less than 26 miles from the major airport were considered for inclusion but upon closer examination we chose to list them as having no airport. A more complicated measure of transportation access based on a gravity model could have resulted in better, or at least different, results.

Black population

The 1970 black population raw count divided by the total population. The data were taken from the US Bureau of the Census, *State and Metropolitan Area data book, 1977*.

Business services

Percentage of employment in accounting, advertising, consulting, research and development laboratories, data processing and computer software services. Source: *State and Metropolitan Area Data Book 1977.* Data were enhanced to reflect 1976 levels of business service employment.

Climate index

Areas were rated on the following climatological characteristics: the number of very hot and very cold months; seasonal temperature variation; the number of heating and cooling degree days; the number of freezing days; the number of zero degree days; the number of 90 degree days. The index was constructed by initially assigning 1,000 points to each place and then deducting points for negative attributes. The data were published in *Places rated almanac* and were collected for the year 1978 (Boyer & Savageau 1981).

Defense spending per capita

Metropolitan defense spending over $10,000 in contract value divided by the population. Data source was the *State and Metropolitan Area data book* and was for 1977.

Educational options

Index, scored 0 to 125, of available options at 2 and 4-year post-secondary educational institutions and professional programs. Data were for 1978; source was *Places rated almanac*.

Fortune 500 headquarters

Raw count of the number of Fortune 500 headquarters located in metropolitan areas. Data were taken from *Fortune* magazine and were collected for 1977–8. Forty-five Fortune 500 companies were not included because their headquarters are not located in major metropolitan areas.

Freeway density

The number of freeway miles in metropolitan areas divided by the metropolitan land area. Data were collected from the US Department of Transportation for the years 1977 and 1978.

Housing price

The average sales price for a home in metropolitan areas in 1976. Data were taken from the annual US Bureau of the Census' survey of the average price of house sales in US metropolitan areas. Data were collected for 1976.

Unemployment rate

Percent unemployment in metropolitan areas. Data source was the *State and Metropolitan Area data book, 1977*.

Unionization rate

Union membership as a percent of non-agricultural and military employment for 1976. Data were only available at the state level, therefore this measure is the percent of unionized labor force for states. Data source was the *Handbook of labor statistics, 1979*.

Wage rate

Average weekly number of hours divided by the average weekly gross wages for manufacturing workers in 1977. Data source was the US Department of Labor, *Employment and earnings state and area series, 1977*. A second source, the *Census of Manufactures, industry statistics*, Table 4, was consulted in those cases where the data were not reported by the Department of Labor. The range across areas varied from approximately $4.00 to $8.00 per hour.

University R&D

Total value of all federal funding to universities in 1977, divided by population. Data source was the National Science Foundation.

APPENDIX 4

Specifying the model

BEFORE ESTIMATING the model, we performed a number of exercises to determine the proper functional form and to detect potential problems with multicollinearity and other sources of misspecification. The first of these was the generation of a skewness measure to test the degree of nonlinearity in the arrays of both exogenous and endogenous variables. The individual metropolitan observations, we suspected, would not produce normal distributions because metropolitan areas are not "normal" in their incidence in the population. Instead, we have a large number of small ones and a very few immense ones. Thus we might expect the distribution of high tech plants and jobs to demonstrate the same skewness, and some of our independent variables to exhibit them as well.

After visually reviewing the resulting scatter plots and calculating kurtosis, we selected a logarithmic form for several of the variables.[1] All of the endogenous variables were transformed in this manner. Of the exogenous variables, five required logging: freeway density, Fortune 500, business services, university research and development funding, and defense spending per capita. The logged variables were then used in linear multiple regression format. The form of the equations estimated was as follows for each of the endogenous variables.

$$P_m = a + b_1 w_m + b_2 u_m + b_3 n_m + b_4 e_m + b_5 h_m + b_6 O_m + b_7 f_m + b_8 A_m + b_9 F_m + b_{10} b_m + b_{11} r_m + b_{12} d_m + b_{13} b_m + e \quad (1)$$

where P_m is the dependent variable, a is the constant, b is the estimate of the parameter, and w_m etc. are the independent variables as shown on pp. 193–5.

The standard assumptions regarding the behavior of the error term were made. In the original model, we had included a number of variables, discussed in the footnotes to the previous section, which turned out to be highly correlated with one another across our set of metropolitan areas.

We chose to eliminate a couple of the offending variables, specifically the arts index and per pupil educational spending (the latter for other reasons cited above as well). As a result, we were able to eliminate almost all of the variables which had simple correlation coefficients in excess of 0.5 (see Table A4.1 for the correlation matrix among these 13 exogenous variables). Only one variable yielded a correlation coefficient higher than 0.5 – airport access, which was modestly correlated with educational options (0.57) and business services (0.63). Nevertheless, the reader should keep in mind that we were not able to remove all of the multicollinearity in the model. To the extent that it remains, the estimates of the parameters may be biased in a manner which cannot be determined.

A related problem is the possibility that the model is misspecified because of missing variables. It is quite likely in an ambitious attempt, as ours is, that certain forces operating on high tech location will be inadequately represented in the model. We are aware of some of them ourselves. We would, for instance, have liked to include a variable capturing the availability of buildable land in the metropolitan area. It proved impossible to find a workable measure. It would also have been desirable to include a variable on tax burden; again, it proved impossible to generate these, at least for true tax incidence rather than apparent rates, at the metropolitan level.[2] The omission of potentially important relationships such as these will of course result in lower levels of explained variation overall, and may also result in parameter misestimation to the extent that any one variable is picking up the contribution of another which is missing.

A final issue is the degree to which spatial autocorrelation could be biasing the estimation results. This occurs when two observations, here individual metropolitan areas, are not truly independent from one another. It is easy to imagine that in contiguous metropolitan areas, such as those in California, the Middle Atlantic and New England areas, the incidence of high tech activity is interrelated. Indeed, the theory we have outlined in Chapter 6 has suggested that one axis of the dispersion process is from core SMSAs to newer adjacent ones. The estimation procedure treats all metropolitan areas as if they were independent observations, whether neighboring or not. Unfortunately, the literature on econometrics to date offers no corrective procedures for this problem, especially when the adjacent units are irregular in size and number and when there are large gaps in the spatial map not covered by any observations at all (i.e. non-metropolitan space). On the other hand,

Table A4.1 The 13 exogenous variables: correlation matrix (Pearson correlation coefficients).

	Airport access	Climate index	Educational options	Unemployment rate	Wage rate	Unionization rate	Housing price	Percent black	Fortune 500 head-quarters	Freeway density	Defense spending	Business services	R&D
airport access		0.1143 (262) P=0.032	0.5701 (264) P=0.000	−0.0958 (259) P=0.062	−0.0668 (262) P=0.141	−0.0502 (264) P=0.208	0.2936 (240) P=0.000	0.1100 (263) P=0.037	0.3936 (87) P=0.000	0.2757 (252) P=0.000	0.1815 (259) P=0.002	0.6372 (259) P=0.000	0.1256 (177) P=0.048
climate index	0.1143 (262) P=0.032		0.2701 (262) P=0.000	0.3152 (257) P=0.000	0.0064 (260) P=0.459	0.1818 (262) P=0.002	0.3529 (238) P=0.000	0.0610 (261) P=0.163	0.1750 (86) P=0.054	0.3035 (250) P=0.000	0.2421 (257) P=0.000	0.1541 (257) P=0.007	−0.0362 (175) P=0.317
educational options	**0.5701 (264) P=0.000**	0.2701 (262) P=0.000		−0.0026 (259) P=0.483	0.0190 (262) P=0.380	0.1156 (264) P=0.030	0.3963 (240) P=0.000	0.1102 (263) P=0.037	0.6163 (87) P=0.000	0.4936 (252) P=0.000	0.2280 (259) P=0.000	0.4974 (259) P=0.000	0.2531 (177) P=0.000
unemployment rate	−0.0958 (259) P=0.062	0.3152 (257) P=0.000	−0.0026 (259) P=0.483		−0.0098 (257) P=0.438	0.1747 (259) P=0.002	0.0097 (236) P=0.441	−0.0235 (258) P=0.354	0.0803 (83) P=0.235	0.0498 (247) P=0.218	0.1026 (254) P=0.051	−0.1620 (254) P=0.005	−0.2003 (172) P=0.004
wage rate	−0.0668 (262) P=0.141	0.0064 (260) P=0.459	0.0190 (262) P=0.380	−0.0098 (257) P=0.438		0.4481 (262) P=0.000	−0.0234 (238) P=0.359	−0.1583 (261) P=0.005	0.1326 (87) P=0.110	0.0862 (250) P=0.087	−0.1483 (257) P=0.009	−0.1033 (258) P=0.049	−0.0761 (177) P=0.157
unionization rate	−0.0502 (264) P=0.208	0.1818 (262) P=0.002	0.1156 (264) P=0.030	0.1747 (259) P=0.002	0.4481 (262) P=0.000		0.1797 (240) P=0.003	−0.4588 (263) P=0.000	0.1699 (87) P=0.058	0.1545 (252) P=0.007	−0.0781 (259) P=0.105	−0.0686 (259) P=0.136	−0.0741 (177) P=0.163
housing price	0.2936 (240) P=0.000	0.3529 (238) P=0.000	0.3963 (240) P=0.000	0.0097 (236) P=0.441	−0.0234 (238) P=0.359	0.1797 (240) P=0.003		−0.1123 (239) P=0.042	0.4162 (85) P=0.000	0.3918 (229) P=0.000	0.1961 (236) P=0.001	0.4811 (236) P=0.000	0.2407 (167) P=0.001

percent black	0.1100	0.0610	0.1102	-0.0235	-0.1583	-0.4588	-0.1123	0.0564	0.0899	0.0938	0.181	0.0354
	(263)	(261)	(263)	(258)	(261)	(263)	(239)	(87)	(251)	(258)	(258)	(176)
	P=0.037	P=0.163	P=0.037	P=0.354	P=0.005	P=0.000	P=0.042	P=0.302	P=0.078	P=0.066	P=0.002	P=0.321
Fortune 500 head-quarters	*0.3936*	0.1750	**0.6163**	0.0803	0.1326	0.1699	*0.4162*		0.5276	0.3020	0.4878	0.2065
	(87)	(86)	**(87)**	(83)	(87)	(87)	(85)		(87)	(86)	(86)	(75)
	P=0.000	P=0.054	**P=0.000**	P=0.235	P=0.110	P=0.058	P=0.000		P=0.000	P=0.002	P=0.000	P=0.038
freeway density	0.2757	*0.3035*	*0.4986*	0.0498	0.0862	0.1545	*0.3918*	0.0899		0.2318	*0.4196*	0.0764
	(252)	(250)	(252)	(247)	(250)	(252)	(229)	(251)		(247)	(248)	(172)
	P=0.000	P=0.000	P=0.000	P=0.218	P=0.087	P=0.007	P=0.000	P=0.078		P=0.000	P=0.000	P=0.160
defense spending	0.1815	0.2421	0.2280	0.1026	-0.1483	-0.0781	0.1961	0.0938	0.2318		0.2505	-0.0425
	(259)	(257)	(259)	(254)	(257)	(259)	(236)	(258)	(247)		(254)	(176)
	P=0.002	P=0.000	P=0.000	P=0.051	P=0.009	P=0.105	P=0.001	P=0.066	P=0.000		P=0.000	P=0.288
business services	**0.6372**	0.1541	*0.4974*	-0.1620	-0.1033	-0.0686	*0.4811*	*0.1812*	*0.4196*	0.2505		0.2387
	(259)	(257)	(259)	(254)	(258)	(259)	(236)	(258)	(248)	(254)		(175)
	P=0.000	P=0.007	P=0.000	P=0.005	P=0.049	P=0.136	P=0.000	P=0.002	P=0.000	P=0.000		P=0.001
R&D	0.1256	-0.0362	0.2531	-0.2003	-0.0761	-0.0741	*0.2407*	0.0354	0.0764	-0.0425	0.2387	
	(177)	(175)	(177)	(172)	(177)	(177)	(167)	(176)	(172)	(176)	(175)	
	P=0.048	P=0.317	P=0.000	P=0.004	P=0.157	P=0.163	P=0.001	P=0.321	P=0.160	P=0.288	P=0.001	

In each set of figures, line 1 is the first order correlation coefficient, line 2 is the number of observations, and line 3 is the *t*-statistic.
Bold figures indicate correlation ≥ 0.50 } shown only for bottom left half of table.
Italic figures indicate correlation ≥ 0.30

dispersion across neighboring metropolitan boundaries is by no means a uniform process and may be expected to respond to the variables we have incorporated, although this response may be stronger by sheer virtue of proximity than it would be if the same metropolitan areas were detached.[3]

Notes

1. All variables assigned a score higher than 2.5 in the skewness measure were logged. The skewness measure computes the differences between the mean and the median and converts it into an index. In this case, our variables were heavily skewed toward the right due to the influence of outlying large metropolitan areas.
2. However, it is also likely that effective tax rates and availability of buildable land are more important in explaining distribution of high tech activity *within* metropolitan areas rather than among them.
3. We considered including a dummy variable to indicate proximity, but decided that this would be too closely correlated with sheer size and apt to reflect simply megalopolis status.

APPENDIX 5

Disaggregating the regression analysis by SMSA size classes

BECAUSE OUR DATA set was so extensive, with 264 observations constituting an entire population, rather than a sample of SMSAs, it proved possible to disaggregate by urban size categories with negligible losses in degrees of freedom. A list of all SMSAs arrayed in descending order by size is contained in Table A5.1 below.

The large SMSA group consisted of all those areas with a population in excess of 1 million in 1977, a group of 37 metro areas headed by the New York SMSA and encompassing those down through San Antonio, Texas. The second group contained those metropolitan areas with population in excess of 500,000 but less than 1 million, accounting for 38 areas altogether headed by Rochester and Sacramento and including down through Flint, Michigan. The final group consisted of all metropolitan areas with 1977 populations below 500,000, a large and disparate group of 189 areas. The largest of these were places like Long Branch–Asbury Park, New Jersey; Raleigh–Durham, North Carolina; West Palm Beach, Florida; and Austin, Texas. The smallest, with populations less than 80,000, included Lawrence, Kansas; Midland, Texas; and Ownesburg, Kentucky.

Large metropolitan areas

Among the large metropolitan areas, only Fortune 500 headquarters could be confirmed as a strong contributor in explaining the distribution of 1977 plants and jobs.[1] The parameters generated for the other variables yielded the same signs as those mentioned above for the entire set of metros, with three interesting exceptions. Wage rates had, for the first time, a negative relationship with high tech activity, suggesting that this variable might have been historically important in siting plants *among* large metropolitan areas. The housing price variable exhibited a perverse, positive sign for this set, suggesting that high housing prices may be a result, rather than a cause, of high tech locational choice. Finally, and most strikingly, business services appeared to be negatively related to variations in high tech activity. In other words, historically, as the theory prescribes, the presence of corporate headquarters and related innovative activities were more important in building big city high tech complexes than was the presence of business services *per se*.

Factors earmarked as significant in recent high tech change for big cities differ markedly from those just discussed. Fortune 500 headquarters is no longer a significant factor; indeed, it takes on a negative value in the jobs equation. The parameters for business services are negative in the plants case and positive in the jobs case, but not significant in either. In other words, major agglomeration factors appeared to be losing their force in the high tech competition among big metro areas. Labor variables, as anticipated, did not play a significant role, except in the case of the unionization variable in the job change regression. Here, for the first time, high rates of unionization do seem to be strongly associated with poorer performance in the job growth category. On the other hand, wage rates display a positive relationship to job growth, strong enough to confirm it as significant if our hypothesis had been that wage rates are *positively* associated with high tech shifts. Amenities variables, with the single exception of climate in the plant regression, are not significant in explaining relative differences among big metropolitan areas.

What is most striking in the case of recent plant and job change among large metropolitan areas are the findings on the defense spending and percent minority variables. Both of these are significant in explaining job and plant changes in the mid-1970s, in the direction hypothesized. This effect was much stronger in explaining recent change than it was in the historically constructed high tech complexes in these large metros, suggesting that defense spending and minority composition are supplanting agglomeration features as distributors of high tech activity among large metropolitan areas. Indeed, high levels of explained variation for changes in plants ($R^2 = 0.65$) and jobs ($R^2 = 0.51$) appear to be statistically attributable for the most part to the political, socioeconomic, and labor variables, rather than to agglomeration factors.

To summarize, then, large metropolitan areas have participated differentially in job and plant growth. Their relative success or failure to maintain their high tech complexes depended upon their position in a process which downplayed the historically important role of agglomerative factors and increasingly favored places with high defense spending receipts per capita, lower percentages of blacks in the population, and lower levels of unionization in the state's workforce. Wage rates did not appear to be a deterrent to relatively strong gains; rather they were positively associated with both job and plant growth, as were housing prices. These findings suggest that forces associated with dispersion are not confined to redistributing high tech activity between large and small metropolitan areas, but are responsible as well for relative shifts in high tech shares among the nation's largest metropolitan areas.

Medium-sized metropolitan areas

As we move down through the metropolitan area subsets by size, the general level of overall variation explained by the model tends to fall. In the case of medium-sized SMSAs, it remained fairly good, ranging from an R^2 of 0.58 in the case of 1977 plant distribution to a low of $R^2 = 0.44$ for job change in 1972–7. But fewer of the parameters were significant in this set, so that it becomes more difficult to confirm our hypotheses about the determinants of high tech location. For this reason, only those variables which could be confirmed as significant or which exhibited provocative signs changes are discussed in this section; the contrasts among size groups is left to the end.

As in the previous set, medium-sized SMSAs showed a tendency for agglomeration factors to be stronger in explaining the overall distribution of plants and jobs than they were in recent changes in those distributions. The presence of Fortune 500 headquarters was a significant positive factor in both plant and job incidence in 1977. Unlike the previous group, where Fortune 500 headquarters actually became negatively associated with job growth in the mid-1970s period, they still appeared to be positively associated with relative growth among performance medium-sized SMSAs, although the parameter generated was not significant. Business services were also significant contributors to variations in plant siting in 1977; the sign on this variable remained positive for the change regressions, although the significance of the parameter could no longer be confirmed. Other than these agglomeration variables, only percent black could be confirmed as having the association hypothesized in the job and plant change regressions. In other words, among medium-sized

SMSAs as well as large ones, racism appeared to be playing a role in discouraging high tech activity in the 1970s.

While not significant, a number of parameters bore signs in these regressions that differed from those for the entire set. In explaining recent job change, for instance, both wage rates and unionization levels assumed their hypothesized negative relationships, which occurred in few other cases. The parameters for several amenities and access variables assumed perverse signs – climate, freeway access, and airport access were all negatively associated with job growth, while housing prices were again positively related. In a case rare in the entire regression analysis, defense spending assumed a negative sign as well. This is suggestive evidence that labor force and socioeconomic variables are more important to medium-sized places in attracting high tech activity than are the amenities, agglomeration, access, and political variables.

Smaller metropolitan areas

Ironically, agglomeration factors and defense spending appear to have played a more important role historically in the distribution of high tech activity among smaller metropolitan areas ($N = 189$) than they did for medium-sized ones. The presence of Fortune 500 headquarters was a significant factor in explaining 1977 plant (though not job) location.[2] Even more striking, the presence of a well-developed business service sector was a significant factor in both plant and job distributions. These findings bear out the view prominent in geography literature that hosting one or more corporations, rather than branch plants alone, strengthens the local economy's performance immensely. Two other variables were significant in explaining historically evolved high tech arrays among this group. One, defense spending, was significant for explaining variations in both job and plant location. Another, freeway density, was a significant factor in explaining plant sites. Interestingly, the airport access variable produced a negative parameter. Thus we might infer that the type of high tech activities sited in smaller cities produces heavy relatively standardized products reliant on truck transportation.

Overall, what these results suggest is that the ability of a smaller metropolitan area to garner a share of high tech growth historically required the presence of access to the interstate freeway system, a business services complex, preferably with at least one Fortune 500 headquarters, and a chunk of defense spending. Without these, attractive wage rates